SpringerBriefs in Electrical and Computer Engineering

Computational Electromagnetics

Series editor

Rakesh Mohan Jha, Bangalore, India

More information about this series at http://www.springer.com/series/13885

Hema Singh · N. Bala Ankaiah
Rakesh Mohan Jha

Active Cancellation
of Probing in Linear Dipole
Phased Array

 Springer

Hema Singh
Centre for Electromagnetics
CSIR-National Aerospace Laboratories
Bangalore, Karnataka
India

Rakesh Mohan Jha
Centre for Electromagnetics
CSIR-National Aerospace Laboratories
Bangalore, Karnataka
India

N. Bala Ankaiah
Centre for Electromagnetics
CSIR-National Aerospace Laboratories
Bangalore, Karnataka
India

ISSN 2191-8112   ISSN 2191-8120   (electronic)
SpringerBriefs in Electrical and Computer Engineering
ISSN 2365-6239   ISSN 2365-6247   (electronic)
SpringerBriefs in Computational Electromagnetics
ISBN 978-981-287-828-1   ISBN 978-981-287-829-8   (eBook)
DOI 10.1007/978-981-287-829-8

Library of Congress Control Number: 2015947799

Springer Singapore Heidelberg New York Dordrecht London

Printed on acid-free paper

Springer Science+Business Media Singapore Pte Ltd. is part of Springer Science+Business Media
(www.springer.com)

To Professor R. Narasimha

In Memory of Dr. Rakesh Mohan Jha
Great scientist, mentor, and excellent
human being

Dr. Rakesh Mohan Jha was a brilliant contributor to science, a wonderful human being, and a great mentor and friend to all of us associated with this book. With a heavy heart we mourn his sudden and untimely demise and dedicate this book to his memory.

Preface

In phased arrays, the generation of adapted pattern according to the signal scenario requires an efficient adaptive algorithm. The phased array along with feed network and signal processing module adapts its beam pattern in a given signal environment. In an arbitrary signal scenario, there may be multiple friendly and hostile sources trying to look into the array. The antenna array is expected to maintain sufficient gain toward each of the desired source(s) and at the same time suppress each probing source. This will cancel transmission of any signal towards each of the hostile probing sources and leads to active cancellation.

In this book, the modified improved LMS algorithm is employed for weight adaptation for generation of beam pattern for multiple signal environments. The performance of dipole phased array is demonstrated in terms of fast convergence, output noise power, and output SINR. The mutual coupling effect and the role of edge elements are taken into account. It is shown that dipole array is able to maintain multilobe beamforming with accurate and deep nulls toward each probing source. This book would be of interest to engineers and researchers working in the area of phased arrays, adaptive array processing, signal processing, and electromagnetics.

Hema Singh
N. Bala Ankaiah
Rakesh Mohan Jha

Acknowledgments

We would like to thank Mr. Shyam Chetty, Director, CSIR-National Aerospace Laboratories, Bangalore for his permission and support to write this SpringerBrief.

We would also like to acknowledge valuable suggestions from our colleagues at the Centre for Electromagnetics, Dr. R.U. Nair, Dr. Shiv Narayan, Dr. Balamati Choudhury, and Mr. K.S. Venu during the course of writing this book. We express our sincere thanks to Mr. Harish S. Rawat, Ms. Neethu P.S., Mr. Umesh V. Sharma, and Mr. Bala Ankaiah, the project staff at the Centre for Electromagnetics, for their consistent support during the preparation of this book.

But for the concerted support and encouragement from Springer, especially the efforts of Suvira Srivastav, Associate Director, and Swati Meherishi, Senior Editor, Applied Sciences & Engineering, it would not have been possible to bring out this book within such a short span of time. We very much appreciate the continued support by Ms. Kamiya Khatter and Ms. Aparajita Singh of Springer toward bringing out this brief.

Contents

About the Authors

Dr. Hema Singh is currently working as Senior Scientist in Centre for Electromagnetics of CSIR-National Aerospace Laboratories, Bangalore, India. Earlier, she was Lecturer in EEE, BITS, Pilani, India during 2001–2004. She obtained her Ph.D. degree in Electronics Engineering from IIT-BHU, Varanasi India in 2000. Her active area of research is Computational Electromagnetics for Aerospace Applications. More specifically, the topics she has contributed to, are GTD/UTD, EM analysis of propagation in an indoor environment, phased arrays, conformal antennas, radar cross section (RCS) studies including Active RCS Reduction. She received Best Woman Scientist Award in CSIR-NAL, Bangalore for period of 2007–2008 for her contribution in the areas of phased antenna array, adaptive arrays, and active RCS reduction. Dr. Singh has co-authored one book, one book chapter, and over 120 scientific research papers and technical reports.

Mr. N. Bala Ankaiah obtained B.Tech. (ECE) from Acharya Nagarjuna University, Andhra Pradesh, India. While he was a Project Engineer at the Centre for Electromagnetics of CSIR-National Aerospace Laboratories, Bangalore, India, he worked on phased antenna arrays and development of adaptive algorithms.

Dr. Rakesh Mohan Jha was Chief Scientist & Head, Centre for Electromagnetics, CSIR-National Aerospace Laboratories, Bangalore. Dr. Jha obtained a dual degree in BE (Hons.) EEE and M.Sc. (Hons.) Physics from BITS, Pilani (Raj.) India, in 1982. He obtained his Ph.D. (Engg.) degree from Department of Aerospace Engineering of Indian Institute of Science, Bangalore in 1989, in the area of computational electromagnetics for aerospace applications. Dr. Jha was a SERC (UK) Visiting Post-Doctoral Research Fellow at University of Oxford, Department of Engineering Science in 1991. He worked as an Alexander von Humboldt Fellow at the Institute for High-Frequency Techniques and Electronics of the University of Karlsruhe, Germany (1992–1993, 1997). He was awarded the Sir C.V. Raman Award for Aerospace Engineering for the Year 1999. Dr. Jha was elected Fellow of INAE in 2010, for his contributions to the EM Applications to Aerospace Engineering. He was also the Fellow of IETE and Distinguished Fellow of ICCES.

Dr. Jha has authored or co-authored several books, and more than five hundred scientific research papers and technical reports. He passed away during the production of this book of a cardiac arrest.

List of Figures

Active Cancellation of Probing in Linear Dipole Phased Array

Abstract In phased arrays, the generation of adapted pattern according to the signal scenario requires an efficient adaptive algorithm. The antenna array is expected to maintain sufficient gain towards each of the desired source and suppress the probing sources. This will cancel the signal transmission towards each of the hostile probing sources and leads to active cancellation. In this book, the modified improved LMS algorithm is employed for weight adaptation of dipole array for the generation of beam pattern in multiple signal environments. The performance of dipole phased array is demonstrated in terms of fast convergence, output noise power and output signal-to-interference-and-noise ratio. The mutual coupling effect and role of edge elements are taken into account. It is shown that dipole array along with an efficient algorithm is able to maintain multilobe beamforming with accurate and deep nulls towards each probing source. This work has application towards active radar cross section (RCS) reduction. This book consists of formulation, algorithm description and result discussion on active cancellation of hostile probing sources in phased antenna array. It includes numerous illustrations demonstrating the theme of the book for different signal environments and array configuration. The concept discussed in this book is simple to understand, even for the students and beginners in the field of phased arrays and adaptive array processing.

Keywords Active cancellation · Adapted pattern · Phased array · Modified improved LMS algorithm · Mutual coupling · Output SINR

1 Introduction

The capability of an efficient adaptive array is to maintain desired signal(s) in a given signal scenario, while placing deep nulls in the probing directions. This is true even when array does not have any a priori information regarding arrival angles, power level, or frequency of the impinging signals. The performance of adaptive arrays depends on the array design parameters significantly (Svendsen and Gupta 2012).

© The Author(s) 2015   1
H. Singh et al., *Active Cancellation of Probing in Linear Dipole Phased Array*,
SpringerBriefs in Computational Electromagnetics,
DOI 10.1007/978-981-287-829-8_1

These parameters include the antenna element radiation pattern (Compton 1982), geometrical configuration (Elliott and Stern 1981a, b; Yuan et al. 2006), mutual coupling (Gupta and Ksienski 1982; Zhang et al. 1987), bandwidth (Gupta et al. 2005), etc. The requirement of limited antenna aperture makes the design compact. The antenna elements are placed close to each other, resulting in greater mutual coupling effect.

The mutual coupling affects the array response i.e. steering vector towards the impinging signals. This results in degradation of direction-of-arrival (DOA) estimation and interference suppression. The performance degradation is even worse for wideband signals scenario (Zhang et al. 1987). This makes essential to analyze the mutual coupling effect and its compensation towards optimized performance of adaptive arrays. The effect of mutual coupling on active cancellation of probing sources has been studied extensively (Hui 2002, 2004; Svendsen and Gupta 2012). The mutual coupling effect is taken into account by including both self and mutual impedance in estimating the received signal. In (Parhizgar et al. 2013), the terminal voltages of antenna elements are decoupled using the mutual impedance matrix (MIM) and direct-data domain (D^3) algorithm (Adve and Sarkar 2000) towards adaptive nulling. The current distribution over the dipole array is determined using method of moments (MoM) for a given signal scenario. The MIM method extracts the open circuited voltages from the measured voltages of antenna terminals, and decouples them. These decoupled open-circuited voltages are used in adaptive algorithm for placing nulls toward each of the probing source. It is reported that this method compensates the mutual coupling effect and accurately place deep nulls along with DOA estimation.

Another method to extract desired signal from a given signal scenario is to employ transformation matrix to the measured voltages at the antenna elements, so as to generate set of induced voltages for uniform linear array of isotropic elements (Kim et al. 2002). Then a D^3 least square algorithm is used to extract information about the desired signals and generate desired adapted pattern. This method is reported to cater to the problems of mutual coupling effect and strong near-field scatterers for uniform and non-uniform arrays.

The mutual coupling compensation can also be achieved by using space-time adaptive processing (STAP) (Griffith and Gupta 2008). It has been reported that in STAP, the nulling performance of adaptive arrays improves with increase in number of antenna elements keeping array aperture fixed (Svendsen and Gupta 2012). This is due to increase in available degrees of freedom for the array. The mutual coupling although increases with the number of array elements, but does not have any adverse consequence on array nulling performance.

Moreover the trend in adaptive processing is to use genetic algorithm based optimization towards the enhancement of adaptive array performance. Tennant et al. (1994) has reported adaptive nulling by perturbations in the position of antenna elements employing a genetic algorithm. Bernardi et al. (2011) has proposed the convex programming based optimization method for the power pattern synthesis and adaptive nulling for circular dipole array. The method includes not only the mutual coupling between the antenna elements buts also the coupling between the antenna elements and the platform.

In this brief, the modified improved LMS algorithm (Singh and Jha 2013) is used for adaptive nulling in linear dipole array. An arbitrary signal environment consisting of multiple narrowband desired signals and probing sources are considered. It is assumed that DoA of the signals is known a priori. Results are discussed for both without and with mutual coupling effect. A mutual impedance matrix consisting of self and mutual impedance is included in the calculating of received field vector at the dipole array. The active cancellation of probing sources in different signal scenario is demonstrated by comparing the adapted pattern with the quiescent pattern. The performance of dipole array is analyzed on the basis of converged output noise power and signal-to-interference-noise ratio (SINR). It is shown that dipole array is able to maintain multilobe beamforming and place deep nulls in each probing direction.

The modified improved LMS algorithm maintains the Toeplitz structure (Godara 2004) of signal covariance matrix, and hence generates optimum weights for generation of adapted pattern even in case of complicated signal scenario. If the direction of probing signal falls within the mainlobe of the array pattern, the main lobe is highly probable to get distorted (Ganz et al. 1990). In case of proposed modified improved LMS algorithm, this is not the case. Even if the probing source lies within the mainlobe of the array pattern, the optimum weights maintain the mainlobe with deep nulls in the direction of probing source. In other words the communication with the desired source is maintained even when the hostile source attempts to probe from the mainlobe direction. Earlier it has been reported that in such a case, there is a noticeable loss of gain for mainlobe with much higher sidelobe level. In this brief, it is not only the mainlobe but also the sidelobe level which is maintained to an acceptable limit. Moreover the probing source is cancelled effectively. This demonstrates the efficacy of the algorithm employed.

2 Formulation for Adapted Pattern in Dipole Array

The radiation behavior of phased array depends on the design parameters including inter-element spacing, individual element characteristics, operating frequency, geometric configuration, and aperture distribution. The geometric configuration and inter-element spacing controls the mutual coupling effect in between the antenna elements. The mutual coupling effect changes the antenna impedance and hence the received signal vector. This effects the array correlation matrix and hence the optimum weights used for excitation of antenna elements. Thus it is not ambiguous to state that the adapted pattern of phased array for a given signal scenario depends on the mutual coupling effect.

2.1   Steering Vector of Linear Dipole Array

A linear dipole array consists of collection of individual and distinguishable antenna elements placed along a straight line (Fig. 1). The response of linear monopole or dipole array depends on the array size, the total number of antenna elements, inter-element spacing, and the type of amplitude excitation.

For an uniform linear dipole array of N elements is the electric field associated with the incident signal is expressed as

$$E_i(\theta, \phi) = f(\theta, \phi) \sum_{i=1}^{N} I_i e^{-j(kd \cos \theta + \alpha)} \tag{1}$$

where, k is the wave number, d is the inter-element spacing, I_i and α are the amplitude and phase of antenna excitations, respectively, and $f(\theta, \phi)$ is the radiation pattern of the dipole element given by (Balanis 2005)

$$f(\theta, \phi) = \frac{j\eta k I_o l e^{-jkr} \sin \theta [2 \cos(kh \cos \theta)]}{4\pi r} \tag{2}$$

The radiation pattern of the antenna array can be obtained by considering only the magnitude of the above expression (1), i.e.

$$|E_i(\theta, \phi)| = |f(\theta, \phi)| \cdot |S| \tag{3}$$

where, $S = \sum_{i=1}^{N} I_i e^{-j(kd \cos \theta + \alpha)}$ is called steering vector or array manifold of array towards the signal incident at an angle of θ.

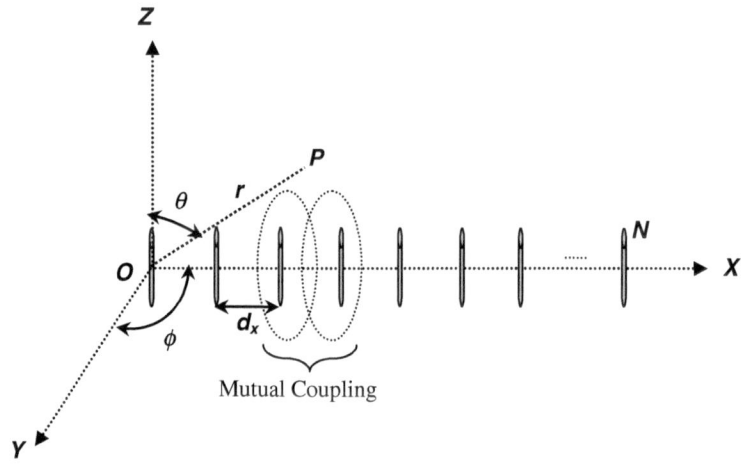

Fig. 1 Schematic diagram of linear dipole array with mutual coupling

2.2 Weight Adaptation in Dipole Array Using Modified Improved LMS Algorithm

The modified improved LMS algorithm is employed to estimate optimum weights for feeding dipole elements. The algorithm involves iterative calculation of antenna weights making use of Toeplitz structure of array correlation matrix (Singh and Jha 2013). The algorithm is capable of maintaining multiple mainlobes towards each of the desired sources. The probing sources are actively cancelled by placing nulls towards them.

The weights for the antenna elements in linear dipole array are iteratively calculated using expressions:

$$W(n+1) = P[W(n) - \mu g(W(n))] + \frac{S}{S^H S} \tag{4}$$

where, n represents the snapshot, μ is the step size, the projection operator, P is given by

$$P = I - \frac{SS^H}{S^H S} \tag{5}$$

I is the identity matrix.

The gradient g in (4) is determined using signal covariance matrix and weights, i.e.

$$g[W(n)] = 2W(n)\tilde{R}(n+1) \tag{6}$$

The correlation matrix is given by

$$\tilde{R}(n+1) = \frac{1}{n+1} \left[\hat{R}(n+1) + n\tilde{R}(n) \right] \tag{7}$$

When the impinging sources (both desired and probing) are uncorrelated, the Toeplitz form of correlation matrix $\hat{R}(n)$, is given by

$$\hat{R}(n) = Z_o^{-1} \begin{bmatrix} \hat{r}_o(n) & \hat{r}_1(n) & \dots & \hat{r}_{(L-1)}(n) \\ \hat{r}_1^*(n) & . & & \\ \vdots & & : & \\ \hat{r}_{(L-1)}^*(n) & & & r_o(n) \end{bmatrix} \tag{8}$$

The impedance matrix Z_o is the impedance matrix.

If multiple desired signals (say p) are simultaneously incident of antenna array, the weight updating equation is modified into

$$W(n+1) = P\left[W(n)\right] - \mu Pg(W(n)) + \frac{S_1}{S_1^H S_1} + \frac{S_2}{S_2^H S_2} + \cdots + \frac{S_p}{S_p^H S_p} \tag{9}$$

Similarly the projection operator will be of the form

$$P = I - \frac{S_1 S_1^H}{S_1^H S_1} - \frac{S_2 S_2^H}{S_2^H S_2} - \cdots - \frac{S_p S_p^H}{S_p^H S_p} \tag{10}$$

where, $S_1, S_2, ..., S_p$ are the steering vectors of array for p desired signals impinging the array from different directions. Once the optimum weights are obtained, the adapted pattern for the given signal scenario is given by

$$\text{Pattern(dB)} = 20 \log_{10}(G) \tag{11}$$

where

$$G = [W(n+1)] \exp[-jd \sin \theta' \cos \phi\prime] \tag{12}$$

2.3   Performance of Linear Dipole Array with Modified Improved LMS Algorithm

The performance of the dipole array is analyzed in terms of output noise power and output SINR. Their converged value w.r.t. the snapshots are obtained as follows:
The output noise power of dipole array is given by

$$P_n = W^H R_n W \tag{13}$$

where R_n is the noise correlation matrix ($N \times N$), expressed as

$$R_n = x_n^H \cdot x_n \tag{14}$$

Here, $x_n = x - x_d$ is the signal vector due to probing sources and thermal noise. x_d is the signal vector due to desired sources, H is the Hermitian of the matrix, and x is the total received signal vector ($1 \times N$), given by

$$x = \sum_{i=1}^{m} p(i) \cdot e^{-j\pi \cos(\theta_i)} \cdot S(\theta_i) \tag{15}$$

where, $m = p + q$, p is the number of desired sources, and q is the number of interfering/probing sources. $S(\theta_i)$ is the array response/steering vector towards the impinging signal at θ_i.

The output signal-to-interference-noise ratio is expressed as

$$\text{SINR} = \frac{W^H R W - W^H R_n W}{W^H R_n W} \tag{16}$$

where, R is the signal correlation matrix,

$$R = x^H \cdot x \tag{17}$$

W is the optimum weight vector calculated using adaptive algorithm for a given signal environment. It should be noted that the mutual coupling effect is included in the impedance matrix and hence the array correlation matrix.

3 Simulation Results: Without Mutual Coupling

This section presents simulation results for linear half-wavelength dipole or quarter-wavelength monopole array. The mutual coupling effect is not included in the computations. The inter-element spacing is taken as $d = 0.484\lambda$. In adapted pattern, the desired sources are shown as green arrows while probing sources are represented by red arrows. Figure 2 shows the output noise power of 16-element

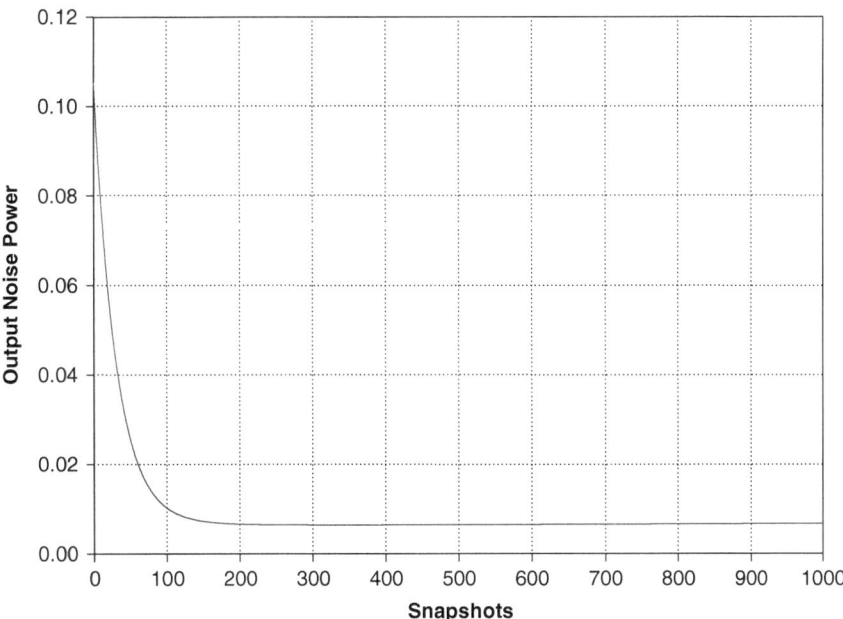

Fig. 2 Output noise power of 16-element linear monopole array. One desired signal (90°; 1) and two probing sources (45°, 135°; 100 each)

linear monopole array with one desired signal (90°; 0 dB) and two probing sources at 45°, 135° with a power ratio of 100 each. The antenna element is taken as quarter-wavelength monopole. The spacing between the monopole array is chosen as 0.484λ. It is apparent that the output noise power converges within few snapshots. Figure 3 shows the corresponding quiescent and adapted patterns. It may be observed that the mainlobe is maintained with deep nulls placed towards each of the probing sources. Moreover the sidelobe level in adapted pattern is also not disturbed significantly.

Next a signal environment consisting of two desired sources and four probing sources is considered. Figure 4 shows the output noise power of 16-element linear monopole array with two desired signals (60°, 100°; 0 dB each) and four probing sources (25°, 45°, 120°, 140°; 1000 each). It can be observed that the converged output noise power level is more than the previous case, i.e. Figure 2, owing to the more number of sources impinging the monopole array.

Next a complicated signal scenario is considered with four desired signals and three probing sources. Figure 5 presents the output noise power of 16-element linear monopole array with four desired signals (50°, 70°, 90°, 110°; 1 each) and three probing sources (35°, 125°, 150°; 1000, 500, 1000). It can be inferred that due to complicated signal environment the output noise power converges after more number of snapshots as compared to Fig. 3. Moreover the performance of monopole array is less in terms of output noise power and depth of nulls in the adapted

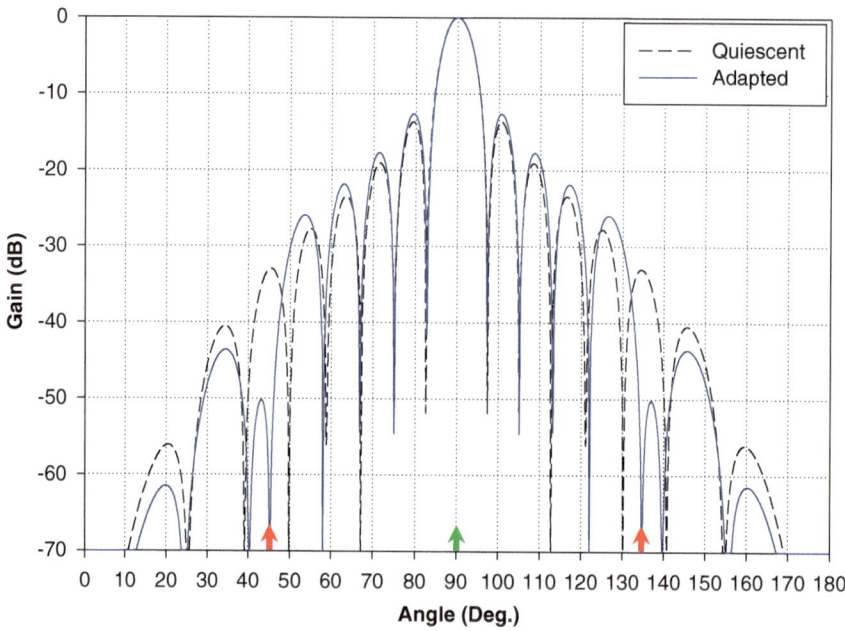

Fig. 3 Adapted beam pattern of 16-element linear monopole array. One desired signal (90°; 1) and two probing sources (45°, 135°; 100 each)

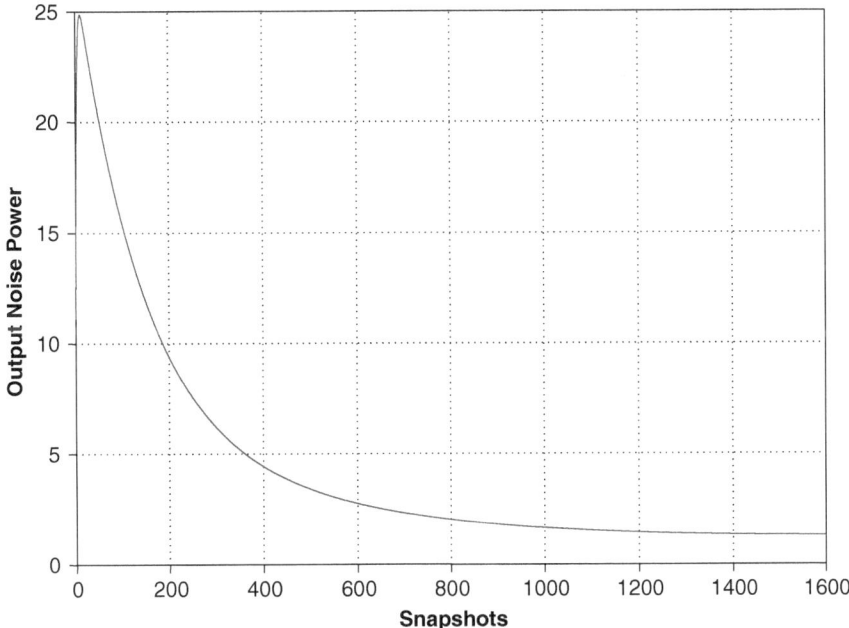

Fig. 4 Output noise power of 16-element linear monopole array. Two desired signals (60°, 100°; 1 each) and four probing sources (25°, 45°, 120°, 140°; 1000 each)

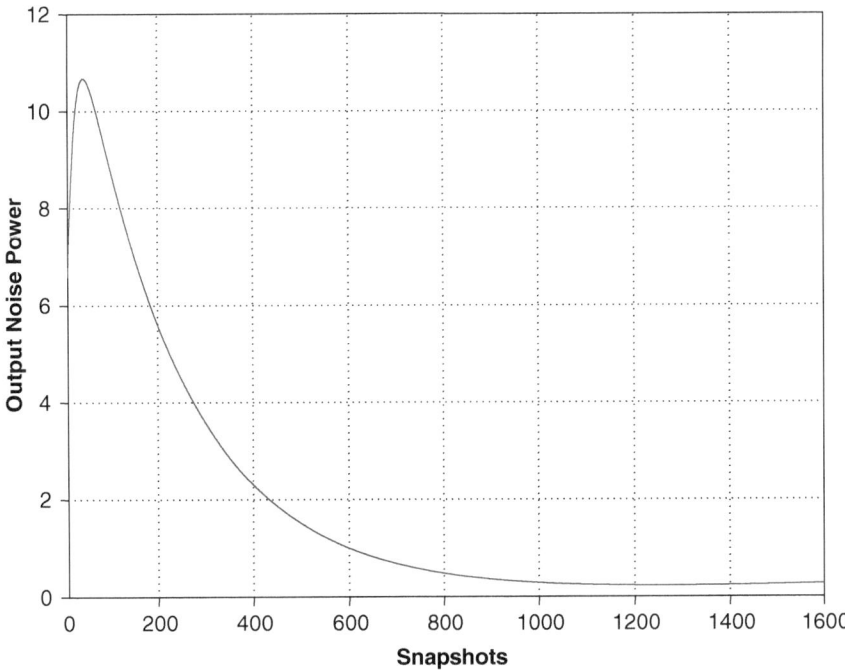

Fig. 5 Output noise power of 16-element linear monopole array. Four desired signals (50°, 70°, 90°, 110°; 1 each) and three probing sources (35°, 125°, 150°; 1000, 500, 1000)

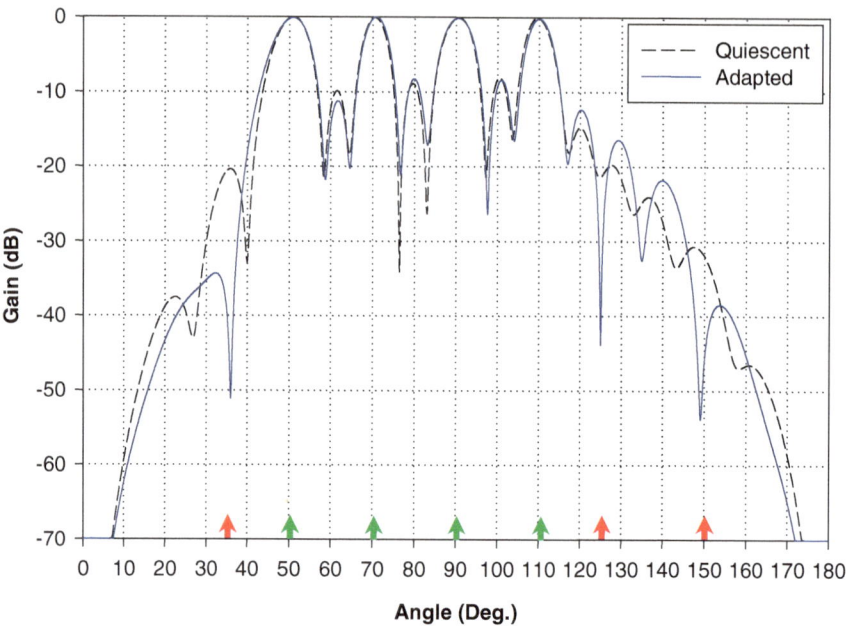

Fig. 6 Adapted beam pattern of 16-element linear monopole array. Four desired signals (50°, 70°, 90°, 110°; 1 each) and three probing sources (35°, 125°, 150°; 1000, 500, 1000)

pattern (Fig. 6). The mainlobe towards each of desired signal is maintained efficiently with each of the probing sources actively cancelled out. This proves the
efficacy of the adaptive algorithm employed for the weight estimation. The dipole
array with the modified improved LMS algorithm is able to cater arbitrary signal
scenario consisting of multiple sources with different power level.

Next, instead of monopole array, half-wavelength dipole array is considered for
analyzing the active cancellation of probing sources. Figure 7 shows the output
noise power of 16-element linear dipole array with one desired signal (90°; 0 dB)
and two probing sources (45°, 135°; 100 each). It can be observed that as in
monopole array, dipole array also performs well with the adaptive algorithm
employed. The output SINR for the same scenario is shown in Fig. 8. The corresponding adapted pattern of 16-element dipole array is shown in Fig. 9. It is
apparent that the array maintains the gain towards the desired source and suppresses
both the probing sources efficiently.

Moreover there is no degradation in the sidelobe level of adapted pattern.

Next two desired signals (40°, 140°; 0 dB) and two probing sources (80°, 100°;
100 each) are assumed to be incident on 16-element linear dipole array. Figure 10
shows the resultant output SINR for the case. The corresponding adapted pattern of
16-element linear dipole array is shown in Fig. 11. It is apparent that the array
maintains the gain towards desired sources and at the same time nulls are placed
towards both the probing sources without any distortion in the mainlobes. Moreover

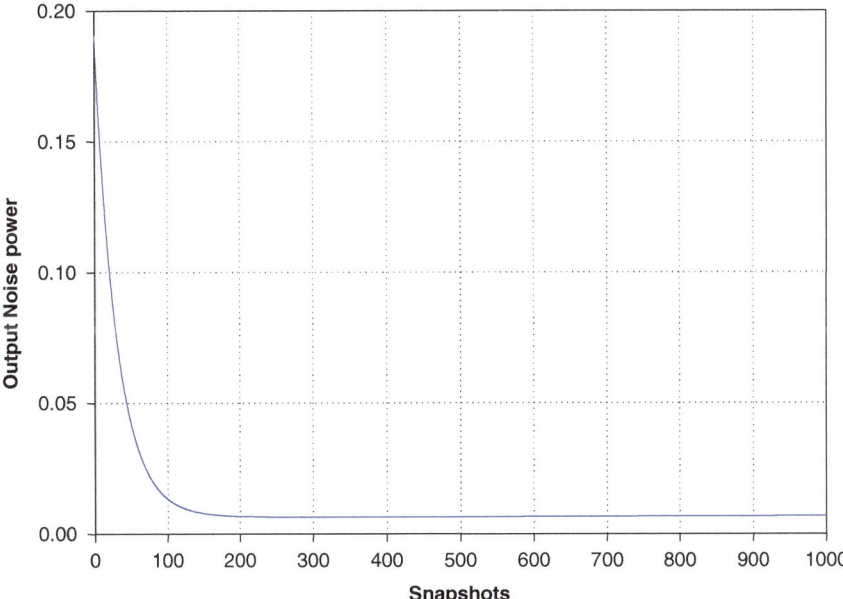

Fig. 7 Output noise power of 16-element linear dipole array. One desired signal (90°; 1) and two probing sources (45°, 135°; 100 each)

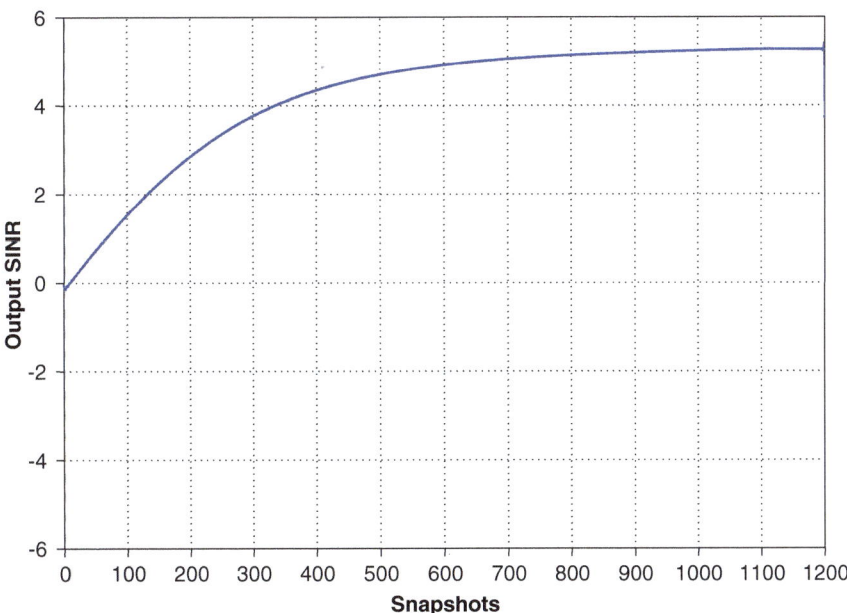

Fig. 8 Output SINR of 16-element linear dipole array. One desired signal (90°; 0 dB) and two probing sources (45°, 135°; 100 each)

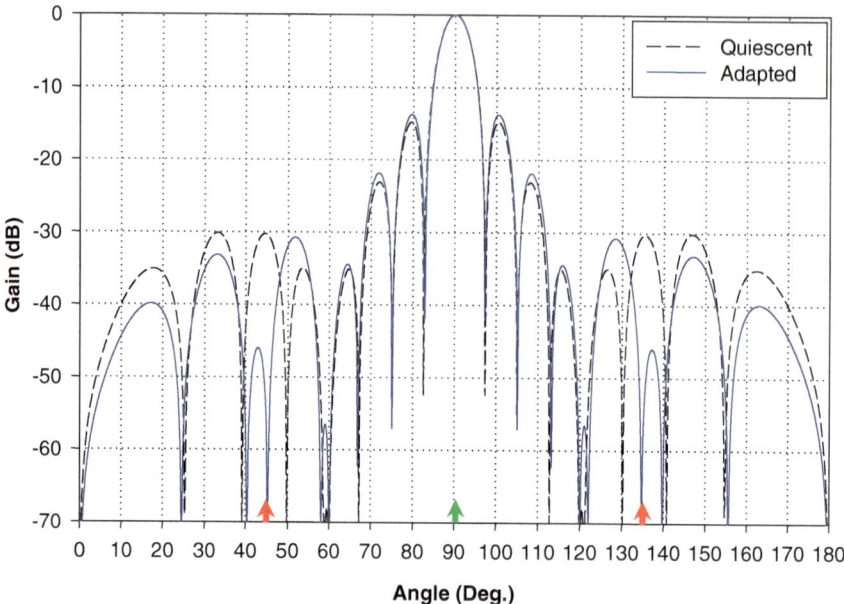

Fig. 9 Adapted pattern of 16-element linear dipole array. One desired signal (90°; 0 dB) and two probing sources (45°, 135°; 100 each)

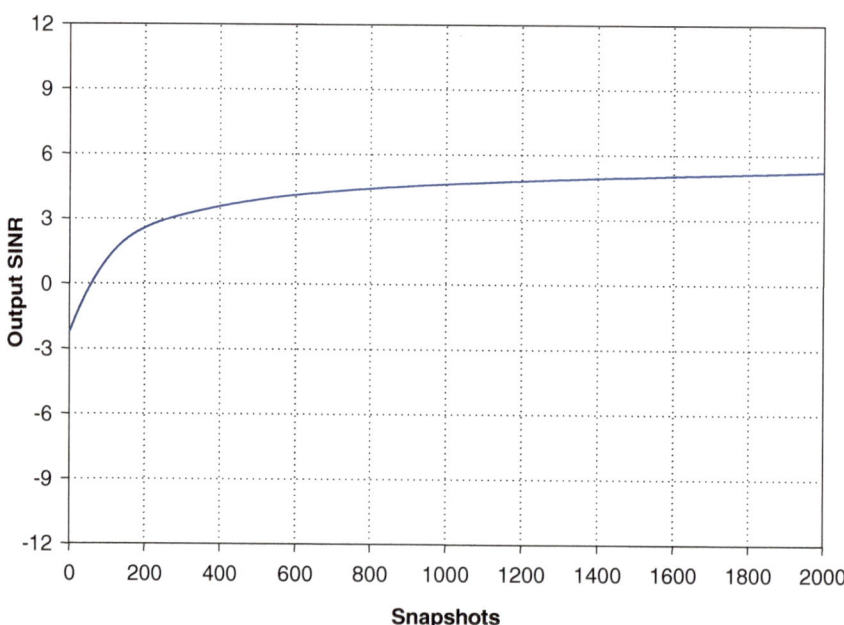

Fig. 10 Output SINR of 16-element linear dipole array. Two desired signals (40°, 140°; 0 dB) and two probing sources (80°, 100°; 100 each)

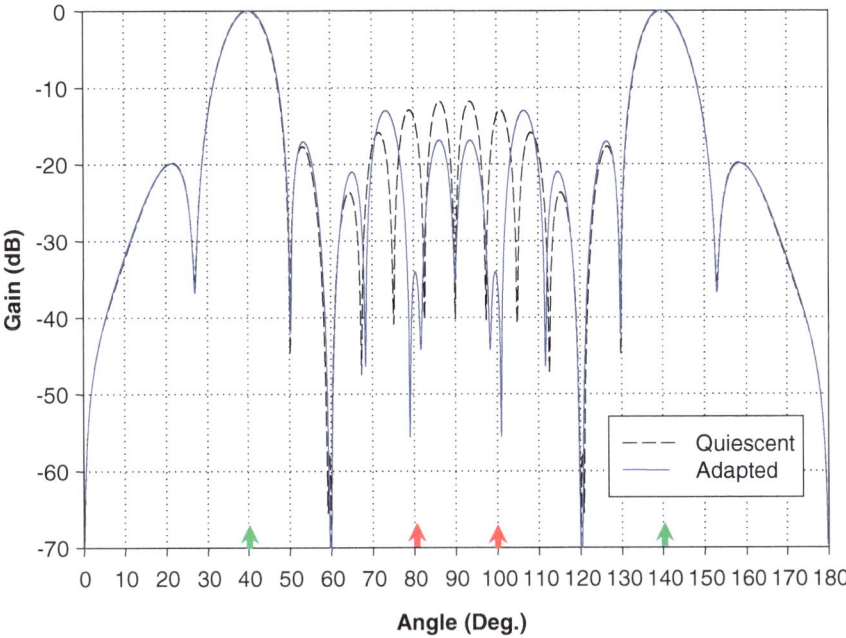

Fig. 11 Adapted pattern 16-element linear dipole array. Two desired signals (40°, 140°; 0 dB) and two probing sources (80°, 100°; 100 each)

the sidelobe level in adapted pattern is lower than in case of quiescent pattern. This proves the capability of the algorithm used for weight adaptation.

Next four desired signals (50°, 70°, 90°, 110°; 0 dB) and three probing sources (35°, 125°, 150°; 1000, 500, 1000) are assumed to impinge a 16-element dipole array. Figure 12 shows the resultant converged output noise power. The convergence rate is however little slower than the earlier cases. This may be due to the complex signal scenario for a linear dipole array.

Now let us consider a probing source (95°; 500) that impinges the array in the direction that lies within the main lobe of the array pattern. Here a 16-element linear monopole array with one desired and two probing sources are considered. The resultant adapted and quiescent patterns are shown in Fig. 13. A sufficiently deep null is placed towards both the probing sources. The probing source that lies within the mainlobe is also suppressed efficiently without much distortion in the mainlobe direction. Moreover the sidelobe distribution remains almost same as in quiescent pattern. This corroborates the efficacy of the modified improved LMS algorithm in active cancellation.

Another signal scenario consisting of four probing sources (30°, 70°, 115°, 135°; 1000 each) and one desired signal (90°; 0 dB) is considered. Figure 14 shows the adapted pattern of a 16-element linear monopole array. Again it is observed that array actively cancel each of probing source maintain the mainlobe towards the desired signal.

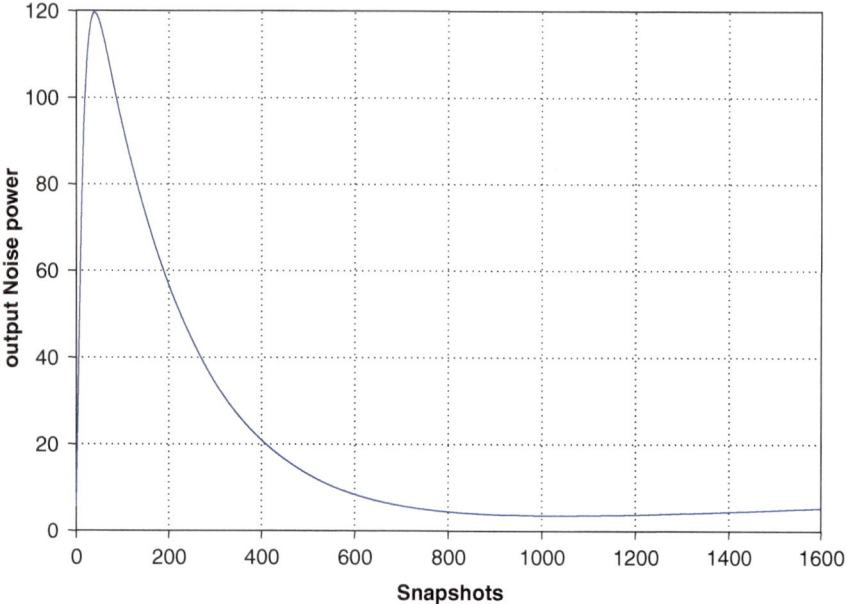

Fig. 12 Output noise power of 16-element linear dipole array. Four desired signals (50°, 70°, 90°, 110°; 1 each) and three probing sources (35°, 125°, 150°; 1000, 500, 1000)

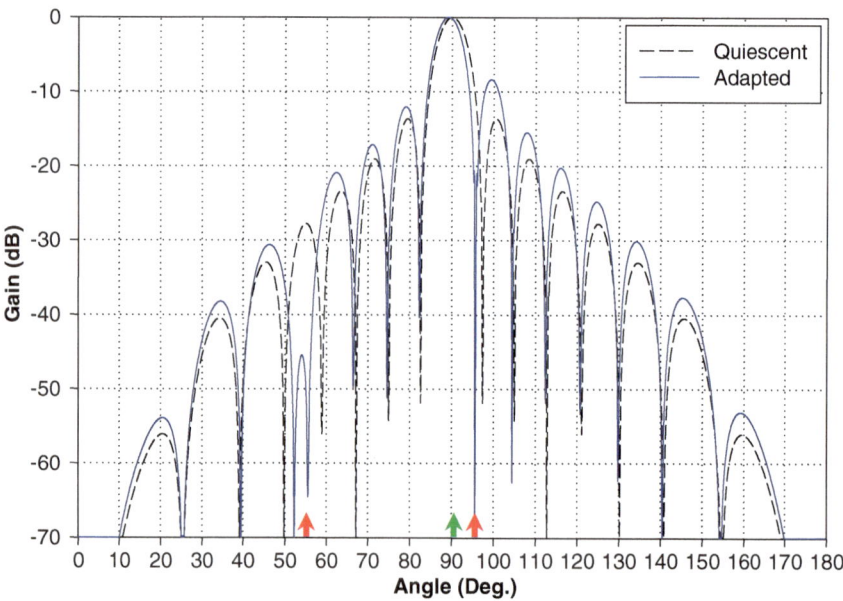

Fig. 13 Adapted beam pattern of 16-element linear monopole array. One desired signal (90°; 1) and two probing sources (55°, 95°; 1000, 500). One probing source falls within the mainlobe

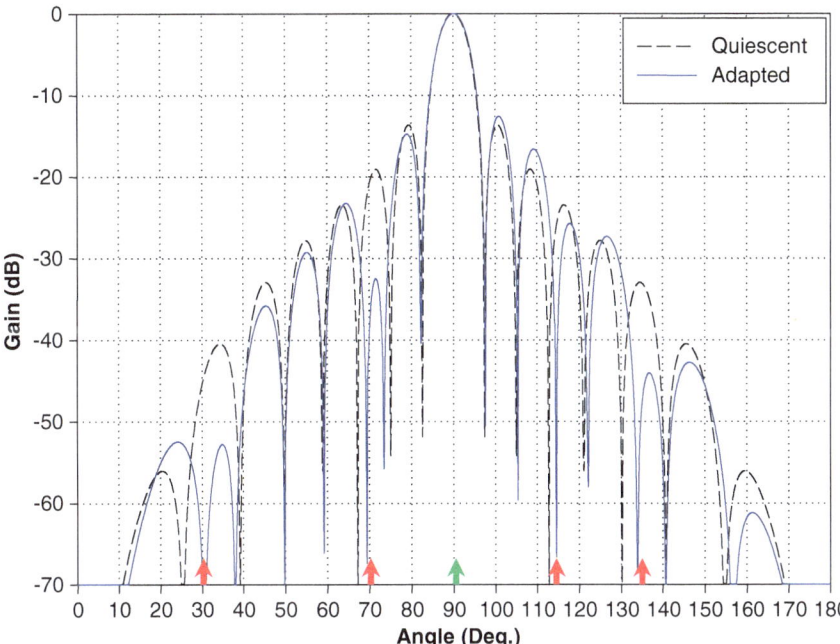

Fig. 14 Adapted beam pattern of 16-element linear monopole array. One desired signal (90°; 1) and four probing sources (30°, 70°, 115°, 135°; 1000)

In order to prove the authenticity of the simulation results, a few more signal environments are considered. Figure 15 presents the adapted pattern for two desired signals (60°, 100°; 0 dB each) and two probing sources (35°, 120°; 100 each). The adapted pattern shows accurate deep nulls towards each probing source, low sidelobe level and two mainlobes without any distortion.

Next for similar signal scenario but one probing source within the mainlobe of the pattern is considered. Figure 16 shows the adapted and quiescent patterns of 16-element linear monopole array for two desired signals and two probing sources. One of the probing source is incident at (65°; 1000), which lies in the main lobe. It can be seen that this probing source is also suppressed efficiently with a negligible shift in the main lobe direction. The number of probing sources is then increased to four (25°, 50°, 120°, 140°; 1000 each) keeping two desired signals (70°, 100°; 1 each).

The resultant adapted pattern is shown in Fig. 17. Again, each of four probing sources is suppressed without any distortion in the two main lobes towards each of the desired signals.

Next the signal environment consisting of three desired signals (60°, 90° and 120°; 0 dB each) and two probing sources (45°, 145°; 100, 1000) is considered. Figure 18 shows that the adapted pattern has deep nulls towards both the probing sources with mainlobes towards each of the desired signals.

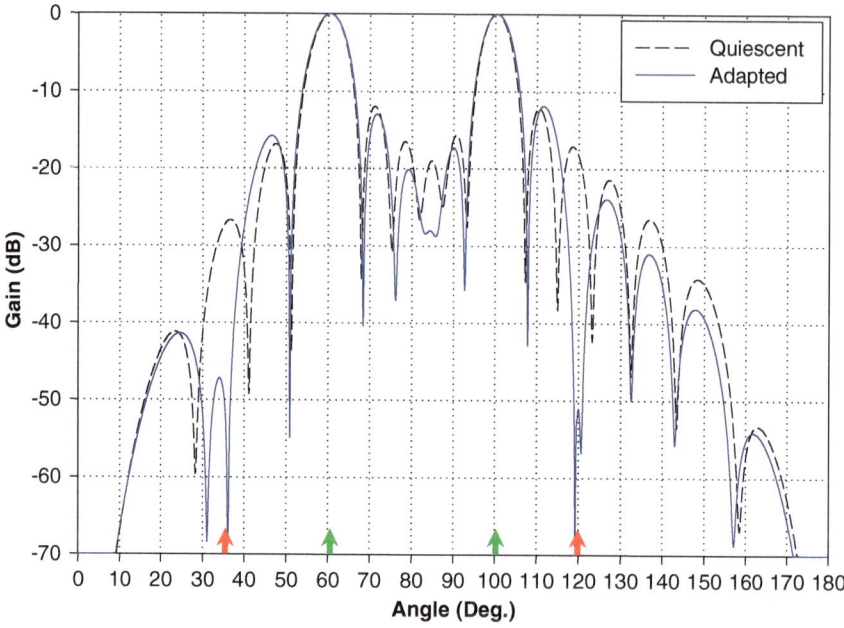

Fig. 15 Adapted beam pattern of 16-element linear monopole array. Two desired signals (60°, 100°; 1 each) and two probing sources (35°, 120°; 100)

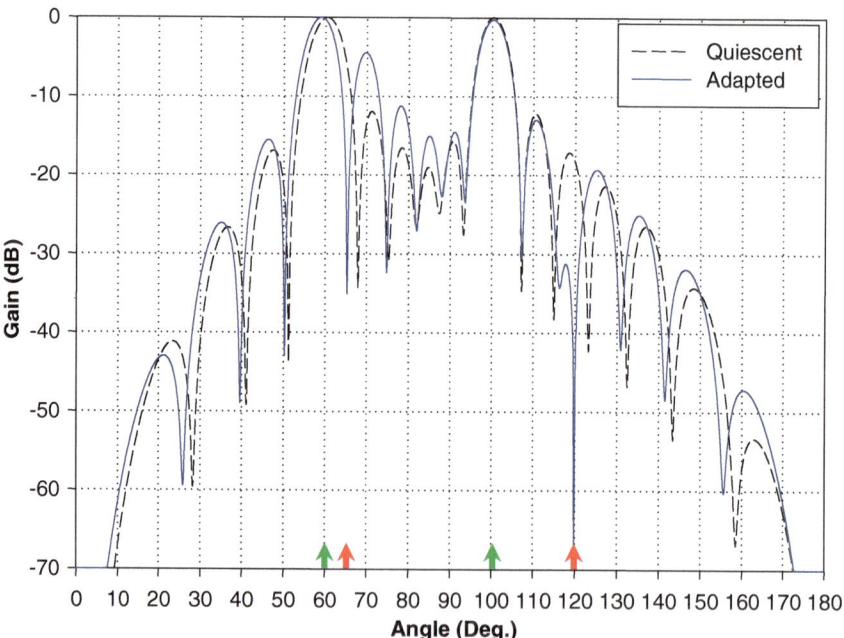

Fig. 16 Adapted beam pattern of 16-element linear monopole array. Two desired signals (60°, 100°; 1) and two probing sources (65°, 120°; 1000)

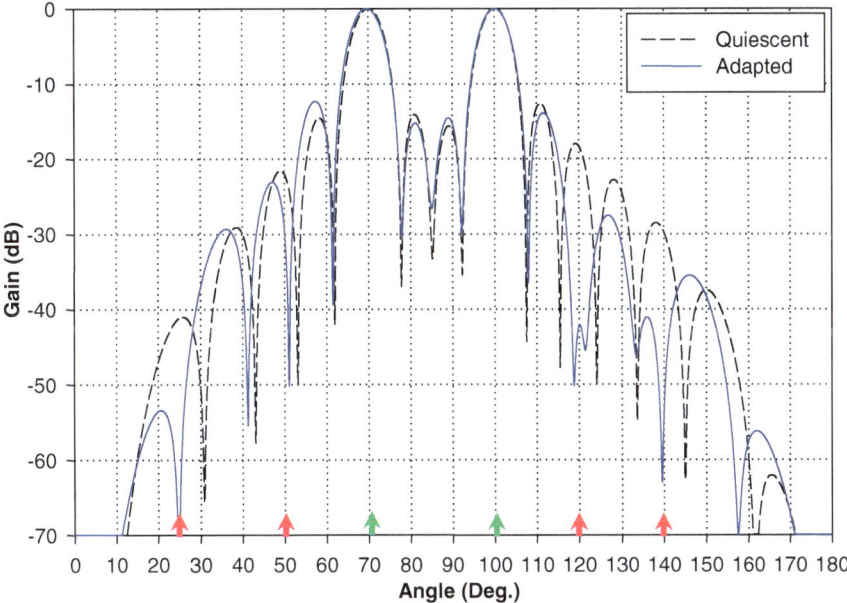

Fig. 17   Adapted pattern of 16-element linear monopole array. Two desired signals (70°, 100°; 1) and four probing sources (25°, 50°, 120°, 140°; 1000)

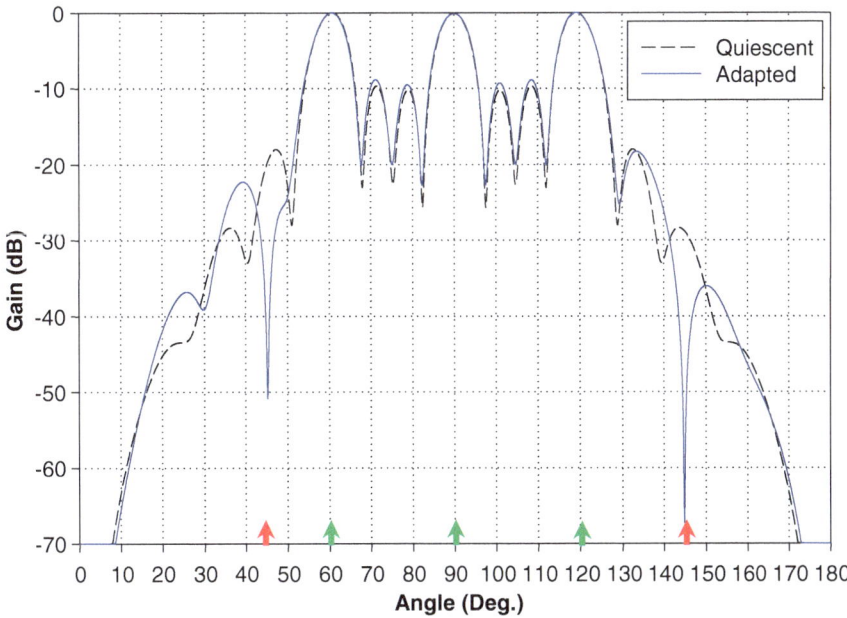

Fig. 18   Adapted pattern of 16-element linear monopole array. Three desired signals (60°, 90°, 120°; 1 each) and two probing sources (45°, 145°; 100, 1000)

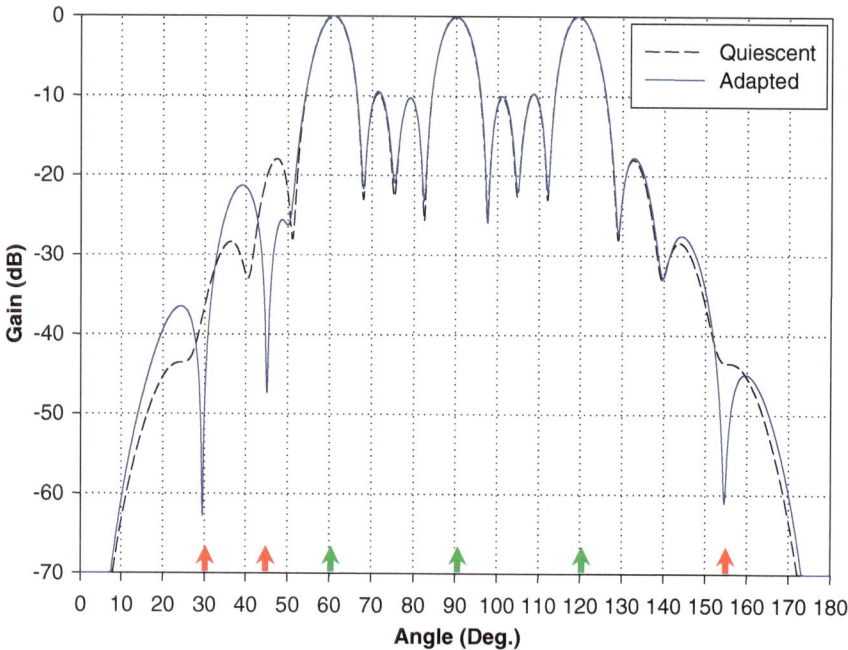

Fig. 19 Adapted pattern of 16-element linear monopole array. Three desired signals (60°, 90°, 120°; 1) and three probing sources (30°, 45°, 155°; 1000)

Figure 19 shows the adapted pattern for three desired signals (60°, 90° and 120°; 0 dB each) and three probing sources (30°, 45°, 155°; 1000 each). It can be observed the nulls placement is accurate and the mainlobe are maintained efficiently. Figure 20 presents the adapted pattern for similar signal scenario consisting of three desired signals (50°, 70°, 90°; 1 each) and three probing sources (115°, 135°, 150°; 800, 1000, 900). The desired signals are assumed to be on one side and the probing sources on the other side of the pattern.

Next a case of four desired signals (50°, 70°, 90°, 110°; 0 dB) and two probing sources (35°, 135°; 1000, 900) is shown. Figure 21 shows the adapted pattern for the case. It can be observed that even four desired signals are maintained with accurate nulls towards each of the probing sources.

It is apparent that whatever the direction of desired signals or probing sources the array can cater to it, adapting its pattern so that probing sources are actively cancelled and the mainlobe towards each of the desired directions is maintained. The mainlobe and the sidelobe level remain undistorted within the limits.

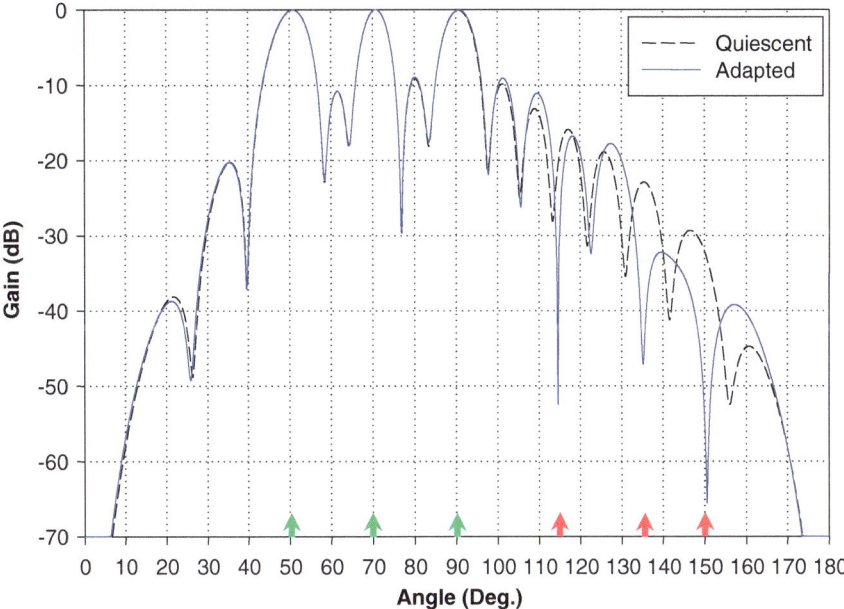

Fig. 20 Adapted pattern of 16-element linear monopole array. Three desired signals (50°, 70°, 90°; 1) and three probing sources (115°, 135°, 150°; 800, 1000, 900)

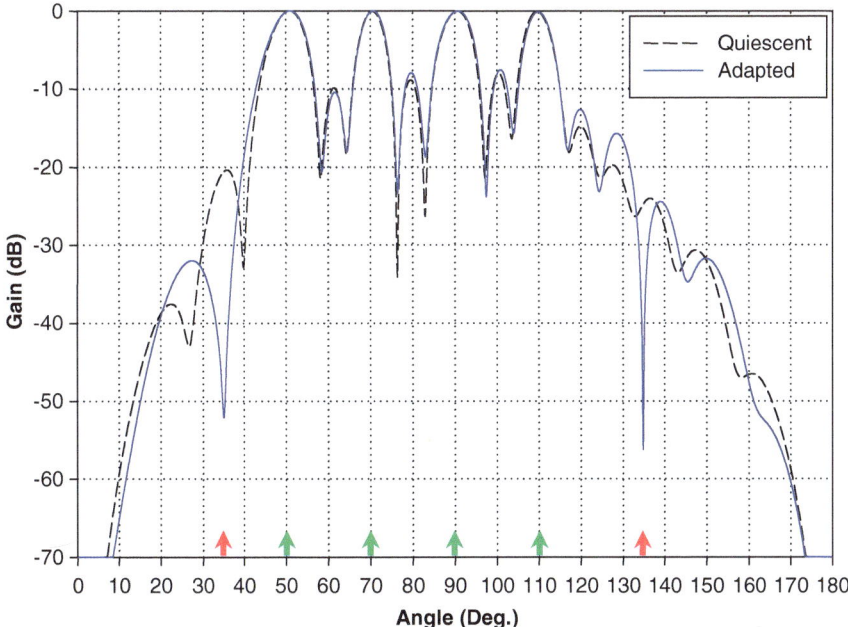

Fig. 21 Adapted pattern of 16-element linear monopole array. Four desired signals (50°, 70°, 90°, 110°; 1) and two probing sources (35°, 135°; 1000, 900)

4   Mutual Coupling Effect in Array Processing

If the antenna elements are placed close to each other, the mutual coupling effect arises. The antenna impedance will thus have both self impedance and mutual impedance components. These components depend on geometric configuration of the antenna array. In this document two configurations, viz. side-by-side and parallel-in-echelon dipole array are considered.

In an N-element linear array, the signal received by each antenna element is weighted and summed to get the array output. The output SINR and output noise power is taken as performance index of adaptive array. The weights used to gain antenna elements need to be adjusted appropriately to maximize the output SINR for a given signal environment.

The impedance matrix of N-element dipole array is given by

$$Z = \begin{bmatrix} Z_{11} & Z_{12} & \cdots & Z_{1N} \\ Z_{21} & Z_{22} & \cdots & Z_{2N} \\ \vdots & \vdots & \ddots & \vdots \\ Z_{N1} & Z_{N2} & \cdots & Z_{NN} \end{bmatrix} \tag{18}$$

where, Z_{ii} is the *self impedance* of the ith element and Z_{ij} is the *mutual impedance* between ith and jth antenna elements. These impedances are complex quantity.

Since each antenna of a N-element array is terminated with the load impedance Z_L, the impedance matrix takes the form

$$Z = \begin{bmatrix} Z_{11} + Z_L & Z_{12} & \cdots & Z_{1N} \\ Z_{21} & Z_{22} + Z_L & \cdots & Z_{2N} \\ \vdots & \vdots & \ddots & \vdots \\ Z_{N1} & Z_{N2} & \cdots & Z_{NN} + Z_L \end{bmatrix} \tag{19}$$

The normalized impedance matrix w.r.t. Z_L is given by

$$Z_{\mathrm{o}} = \begin{bmatrix} 1 + \frac{Z_{11}}{Z_L} & \frac{Z_{12}}{Z_L} & \cdots & \frac{Z_{1N}}{Z_L} \\ \frac{Z_{21}}{Z_L} & 1 + \frac{Z_{22}}{Z_L} & \cdots & \frac{Z_{2N}}{Z_L} \\ \vdots & \vdots & \ddots & \vdots \\ \frac{Z_{N1}}{Z_L} & \frac{Z_{N2}}{Z_L} & \cdots & 1 + \frac{Z_{NN}}{Z_L} \end{bmatrix} \tag{20}$$

In the absence of mutual coupling, impedance matrix becomes a diagonal matrix, and is given by

$$Z_0 = \begin{bmatrix} 1+\frac{Z_{11}}{Z_L} & 0 & \cdots & 0 \\ 0 & 1+\frac{Z_{22}}{Z_L} & \cdots & 0 \\ \vdots & \vdots & \ddots & \vdots \\ 0 & 0 & \cdots & 1+\frac{Z_{NN}}{Z_L} \end{bmatrix} \tag{21}$$

4.1 Side-by-Side Dipole Array

The schematic of a side-by-side dipole array is shown in Fig. 22. The length of the dipole element is l, and d is the spacing between the two adjacent dipole elements.

For a pair of $\lambda/2$ dipoles placed side-by-side, placed at distance d, the self and mutual impedances are given by

$$R_{\text{self}} = \frac{\eta}{2\pi}\left\{C + \ln(kl) - C_i(kl) + \frac{1}{2}\sin(kl)[S_i(2kl) - 2S_i(kl)]\right.$$
$$\left. + \frac{1}{2}\cos(kl)[C + \ln(kl/2) + C_i(2kl) - 2C_i(kl)]\right\} \tag{22}$$

$$X_{\text{self}} = \frac{\eta}{4\pi}\left\{2S_i(kl) + \cos(kl)[2S_i(kl) - S_i(2kl)]\right.$$
$$\left. - \sin(kl)\left[2C_i(kl) - C_i(2kl) - C_i\left(\frac{2ka^2}{l}\right)\right]\right\} \tag{23}$$

$$R_{\text{mutual}} = \frac{\eta}{4\pi}[2C_i(u_0) - C_i(u_1) - C_i(u_2)] \tag{24}$$

$$X_{\text{mutual}} = -\frac{\eta}{4\pi}[2S_i(u_0) - S_i(u_1) - S_i(u_2)] \tag{25}$$

where, $C_i(x)$ and $S_i(x)$ are cosine and sine integrals expressed as (Balanis 2005)

Fig. 22 Side-by-side configuration of dipole antenna of length l

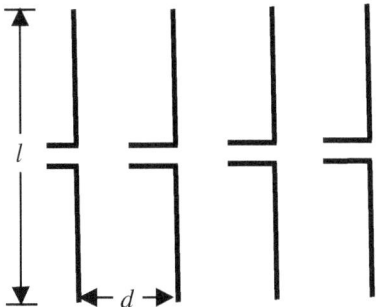

$$S_i(x) = \sum_{k=0}^{\infty} \frac{(-1)^k x^{2k+1}}{(2k+1)(2k+1)!} \tag{26}$$

$$C_i(x) = C + \ln(x) + \sum_{k=1}^{\infty} (-1)^k \frac{x^{2k}}{2k(2k)!} \tag{27}$$

where, $u_0 = kd$; $u_1 = k(\sqrt{d^2 + l^2} + l)$; $u_2 = k(\sqrt{d^2 + l^2} - l)$
$C = 0.5772156649$, is the Euler's constant.

4.2 Parallel-in-Echelon Dipole Array

The schematic of a typical parallel-in-echelon dipole array is shown in Fig. 23. The height H of one of the elements is zero w.r.t. the reference plane while the other element is at height H.

For a pair of $\lambda/2$ dipoles placed in parallel-in-echelon configuration, spaced at distance d, the self and mutual impedances are given by

$$R_{\text{self}} = \frac{\eta}{2\pi} \left\{ C + \ln(kl) - C_i(kl) + \frac{1}{2}\sin(kl)[S_i(2kl) - 2S_i(kl)] \right.$$
$$\left. + \frac{1}{2}\cos(kl)[C + \ln(kl/2) + C_i(2kl) - 2C_i(kl)] \right\} \tag{28}$$

$$X_{\text{self}} = \frac{\eta}{4\pi} \left\{ 2S_i(kl) + \cos(kl)[2S_i(kl) - S_i(2kl)] \right.$$
$$\left. - \sin(kl)\left[2C_i(kl) - C_i(2kl) - C_i\left(\frac{2ka^2}{l}\right) \right] \right\} \tag{29}$$

Fig. 23 Schematic of parallel-in-echelon configuration of a dipole array

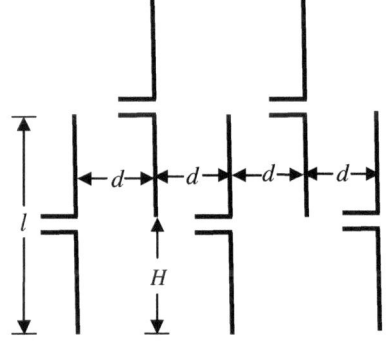

$$R_{\mathrm{mutual}} = -\frac{\eta}{8\pi}\cos(w_0)\begin{bmatrix} -2C_i(w_1) - 2C_i(w_1') + C_i(w_2) + \\ C_i(w_2') + C_i(w_3) + C_i(w_3') \end{bmatrix}$$
$$+\frac{\eta}{8\pi}\sin(w_0)\begin{bmatrix} -2S_i(w_1) - 2S_i(w_1') + S_i(w_2) \\ +S_i(w_2') + S_i(w_3) + S_i(w_3') \end{bmatrix} \tag{30}$$

$$X_{\mathrm{mutual}} = -\frac{\eta}{8\pi}\cos(w_0)\begin{bmatrix} -2S_i(w_1) + 2S_i(w_1') - S_i(w_2) \\ -S_i(w_2') - S_i(w_3) - S_i(w_3') \end{bmatrix}$$
$$+\frac{\eta}{8\pi}\sin(w_0)\begin{bmatrix} -2C_i(w_1) - 2C_i(w_1') - C_i(w_2) \\ +C_i(w_2') - C_i(w_3) + C_i(w_3') \end{bmatrix} \tag{31}$$

where, $w_0 = kH$; $w_1 = k\left(\sqrt{d^2 + H^2} + H\right)$; $w_1' = k\left(\sqrt{d^2 + H^2} - H\right)$;
$w_2 = k\left[\sqrt{d^2 + (H-l)^2} + (H-l)\right]$; $w_2' = k\left[\sqrt{d^2 + (H-l)^2} - (H-l)\right]$ $w_3 =$
$k\left[\sqrt{d^2 + (H+l)^2} + (H+l)\right]$; $w_3' = k\left[\sqrt{d^2 + (H+l)^2} - (H+l)\right]$

5   Simulation Results: With Mutual Coupling

Here a 16-element uniform linear dipole array is considered. The dipole length is taken as half-wavelength with $h = 0.25\lambda$. The inter-element spacing is 0.484λ. The results are shown for side-by-side and parallel-in-echelon dipole arrays for different signal environments. In adapted pattern, the desired sources are shown as green arrows while probing sources are represented by red arrows.

5.1   Side-by-Side Configuration

Figure 24 shows the output noise power of 16-element linear dipole array with one desired signal (90°; 1) and two probing sources (45°, 135°; 100 each). It is apparent that the output noise power converges beyond 400 snapshots for both with and without mutual coupling effect. The corresponding output SINR is shown in Fig. 25. As expected the output SINR of the array is lower when mutual coupling effect is included in the calculations.

Figure 26 shows the adapted pattern with mutual coupling and without mutual coupling along with the quiescent pattern. The signal scenario is same as in Figs. 24 and 25. It may be observed that the mainlobe is in the desired direction (90°) and deep nulls are placed towards each probing direction (45°, 135°).

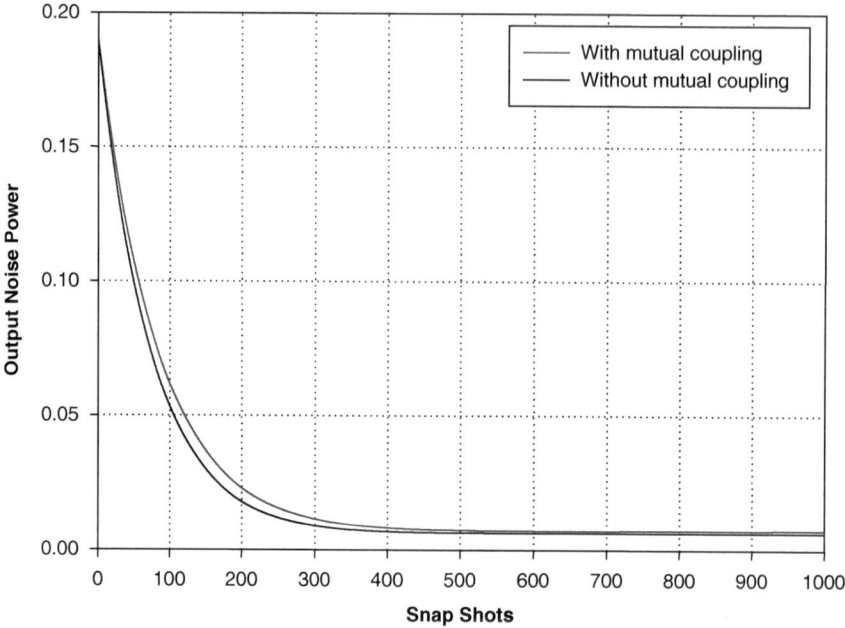

Fig. 24 Output noise power of 16-element linear dipole array. One desired signal (90°; 1) and two probing sources (45°, 135°; 100 each)

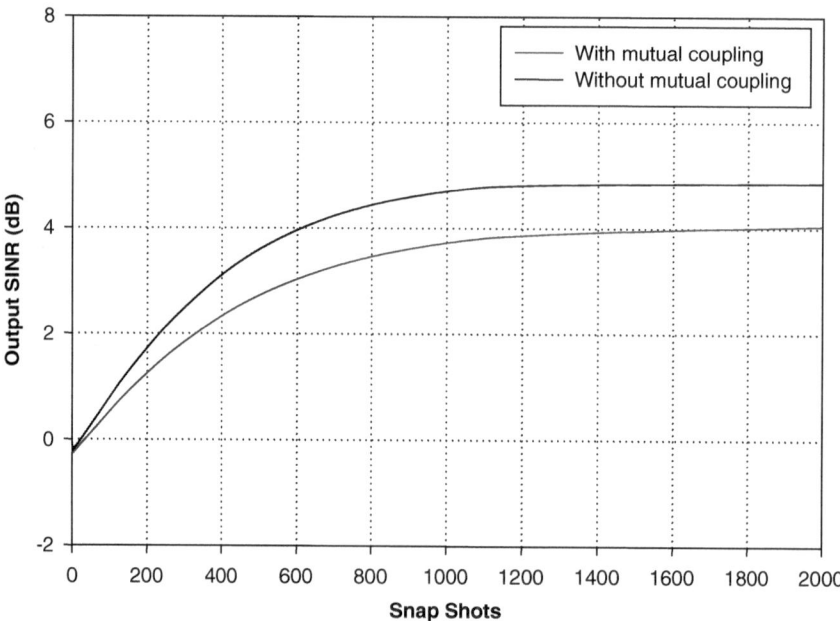

Fig. 25 Output SINR of 16-element linear dipole array. One desired signal (90°; 1) and two probing sources (45°, 135°; 100 each)

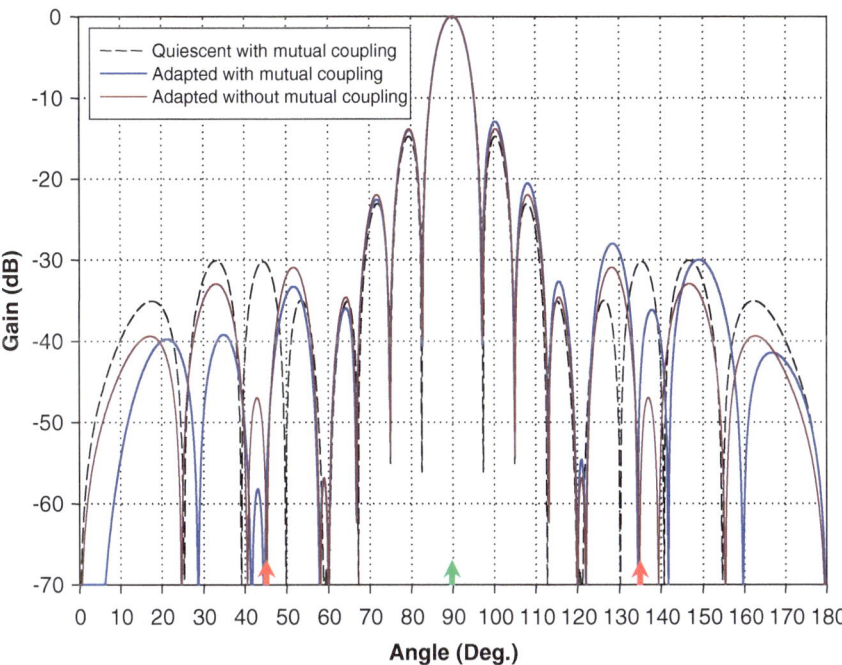

Fig. 26 Adapted pattern of 16-element linear dipole array. One desired signal (90°; 1) and two probing sources (45°, 135°; 100 each)

Next, the inter-element spacing is varied (0.284λ, 0.384λ, 0.484λ) and its effect on adapted pattern is analyzed. Same signal scenario of one desired signal (90°; 1) and two probing sources (45°, 135°; 100 each) is considered. The adapted patterns of 16-element linear dipole array for three inter-element spacings are compared in Fig. 27. It can be observed that in all the three cases of inter-element spacings, the mainlobe towards desired signal is maintained efficiently with accurately placed nulls towards each of the probing sources. However, the broadness of mainlobe towards the desired source increases as the elements are brought closer. Further the depths of the nulls become less for smaller inter-element spacing. This might be due to the dominance of mutual coupling effect when elements are brought closer.

Next the number of probing sources is increased to three. Figure 28 shows the adapted pattern for one desired signal (90°; 1) and three probing sources (35°, 80°, 135°; 100, 100, 200).

It is apparent that the all the three probing sources are actively cancelled with mainlobe towards the desired source. The mutual coupling effect is included in the weight adaptation.

Next two desired signals (40°, 140°; 1 each) and two probing sources (80°, 110°; 1000 each) are assumed to be incident on a 16-element linear dipole array. Figure 29 shows the resultant output noise power with and without mutual coupling. The output noise power although converges but beyond 1000 snapshots.

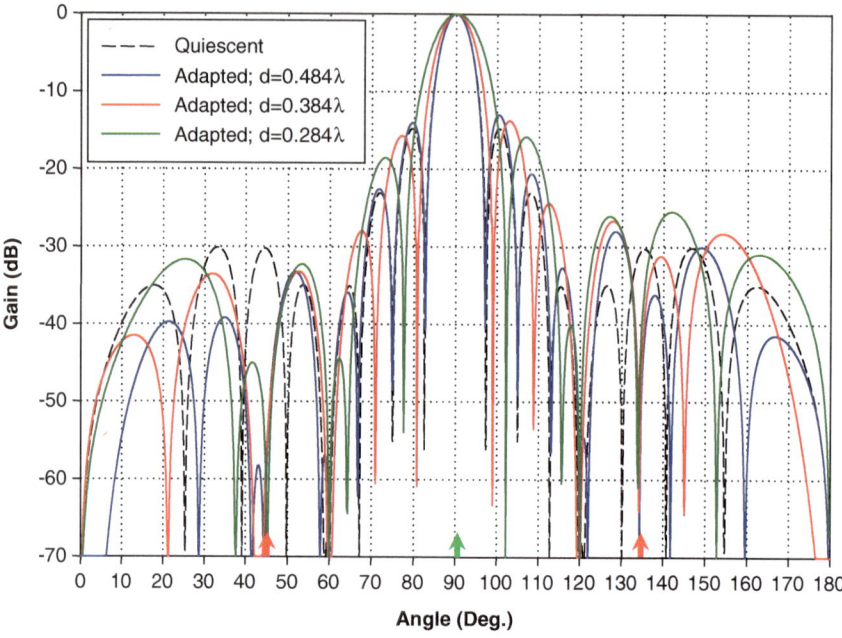

Fig. 27 Effect of inter-element spacing on adapted beam pattern. One desired signal (90°; 1) and two probing sources (45°, 135°; 100 each)

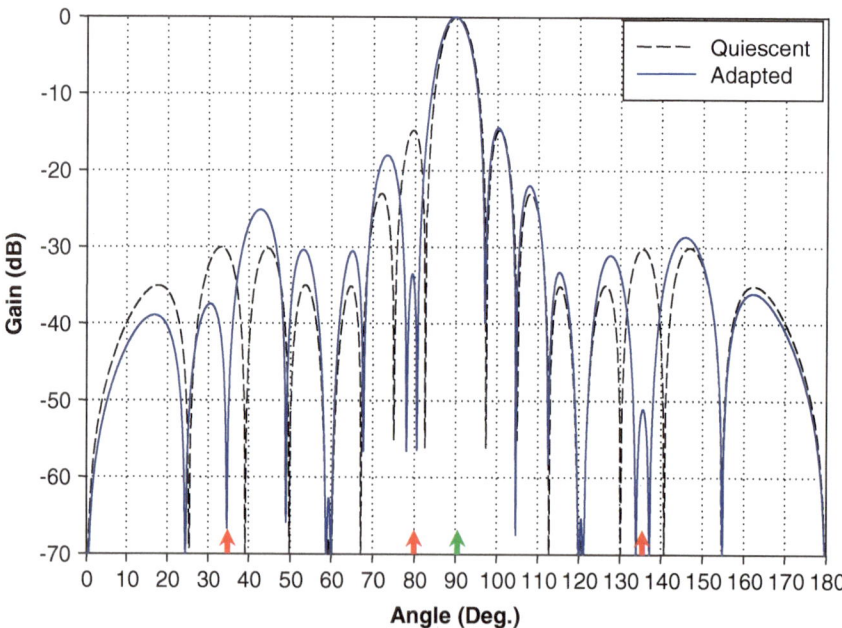

Fig. 28 Adapted beam pattern of 16-element linear dipole array. One desired signal (90°; 1) and three probing sources (35°, 80°, 135°; 100, 100, 200)

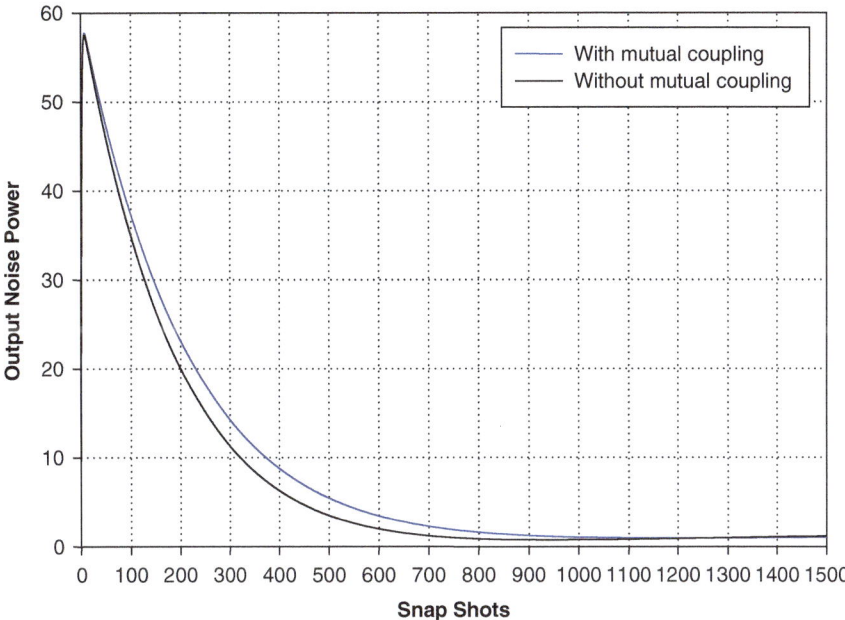

Fig. 29 Output noise power of 16-element linear dipole array. Two desired signals (40°, 140°; 1 each) and two probing sources (80°, 110°; 1000 each)

Since the signal scenario is complex (two desired and two probing) as compared to Fig. 24 (one desired, two probing), the convergence rate of the output noise power is less. The corresponding adapted pattern is shown in Fig. 30. The array maintains the gain towards each of the desired sources and simultaneously placed deep nulls towards both the probing sources. There is no distortion in the mainlobes of adapted pattern.

Next another signal scenario is considered. Figure 31 shows output SINR of 16-element linear dipole array with two desired signals (40°, 90°; 1 each) and two probing sources (110°, 135°; 100 each). As expected, the output SINR of dipole array is lower when mutual coupling is included, as compared to no mutual coupling case. The corresponding adapted pattern of is shown in Fig. 32. It can be observed that the array maintains the gain towards the desired sources and suppresses both the probing sources efficiently.

Next the number of probing sources is increased to three. Figure 33 presents the adapted pattern for two desired signals (40°, 90°; 1 each) and three probing sources (80°, 110°, 150°; 1000 each). The adapted pattern has accurate deep nulls towards each probing source. The mainlobes are maintained towards each of desired directions without any distortion. Moreover the sidelobe level is maintained low throughout the pattern.

Next the number of desired signals is increased to three (40°, 90°, 140°; 1 each). The number of probing sources are two (70°, 110°; 100 each). The resultant adapted

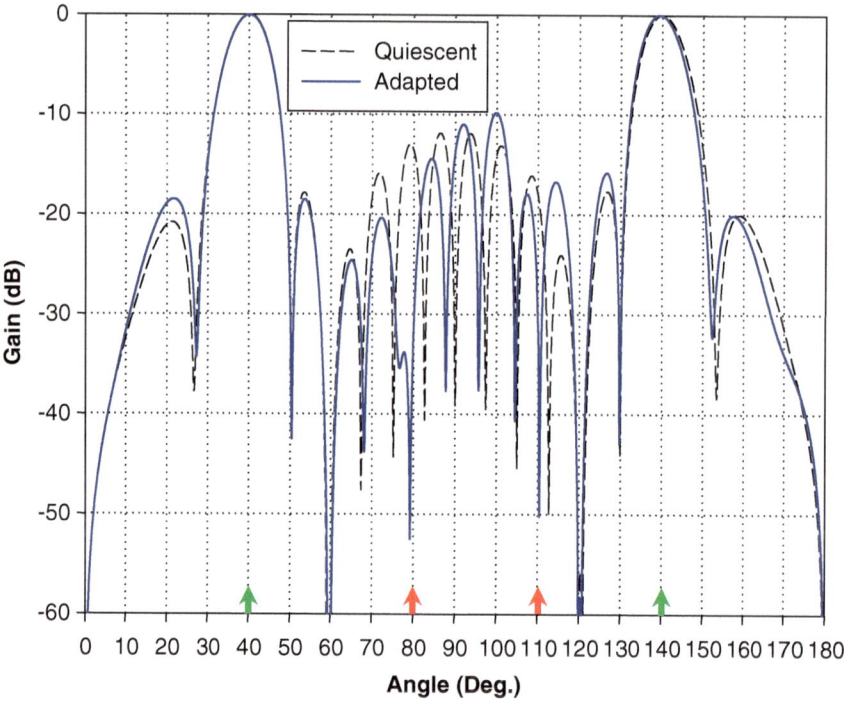

Fig. 30 Adapted beam pattern of 16-element linear dipole array. Two desired signals (40°, 140°; 1 each) and two probing sources (80°, 110°; 1000 each)

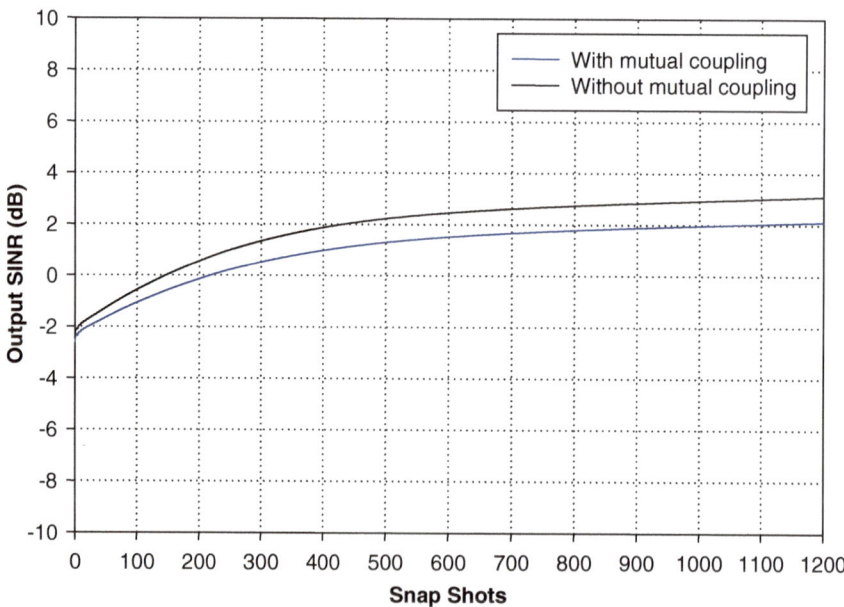

Fig. 31 Output SINR of 16-element linear dipole array. Two desired signals (40°, 90°; 1 each) and two probing sources (110°, 135°; 100 each)

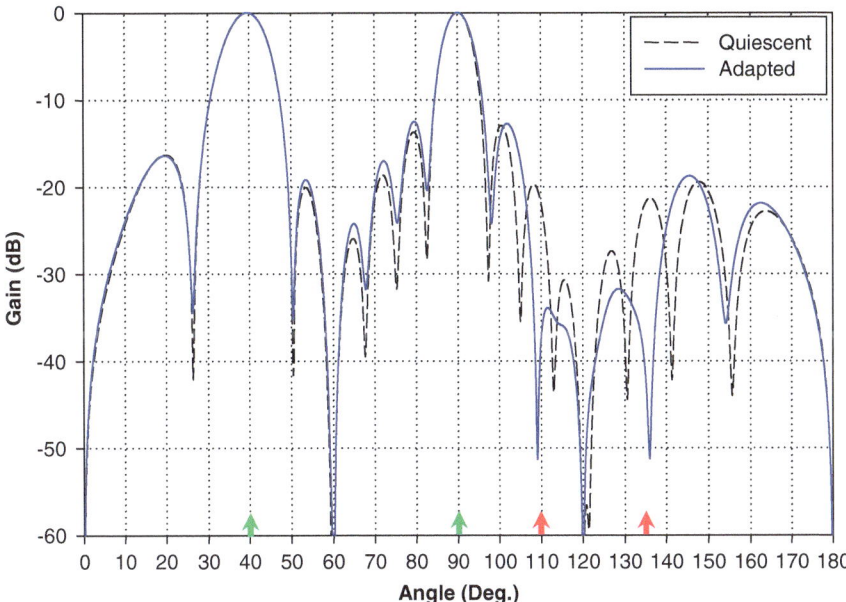

Fig. 32 Adapted beam pattern of 16-element linear dipole array. Two desired signals (40°, 90°; 1) and two probing sources (110°, 135°; 100)

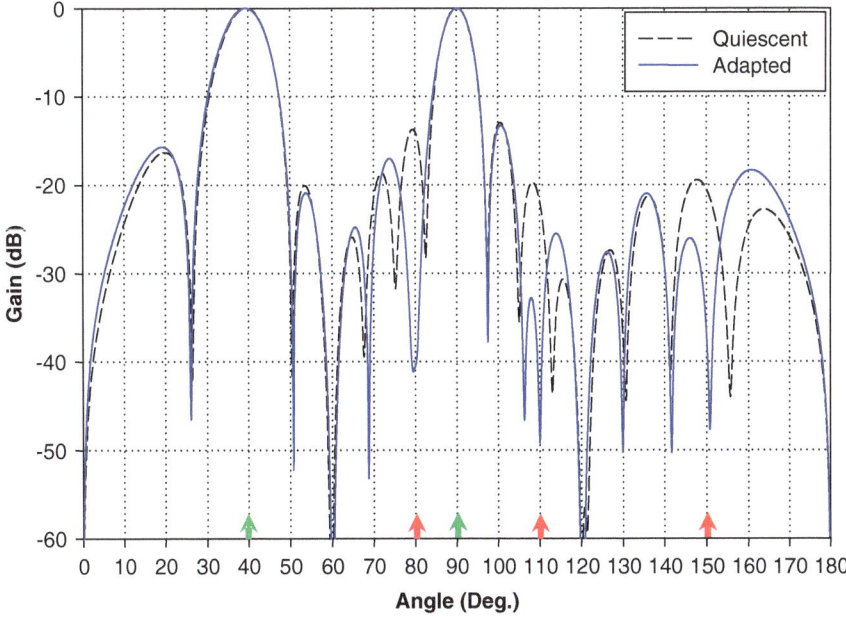

Fig. 33 Adapted beam pattern of 16-element linear dipole array. Two desired signals (40°, 90°; 1) and three probing sources (80°, 110°, 150°; 1000)

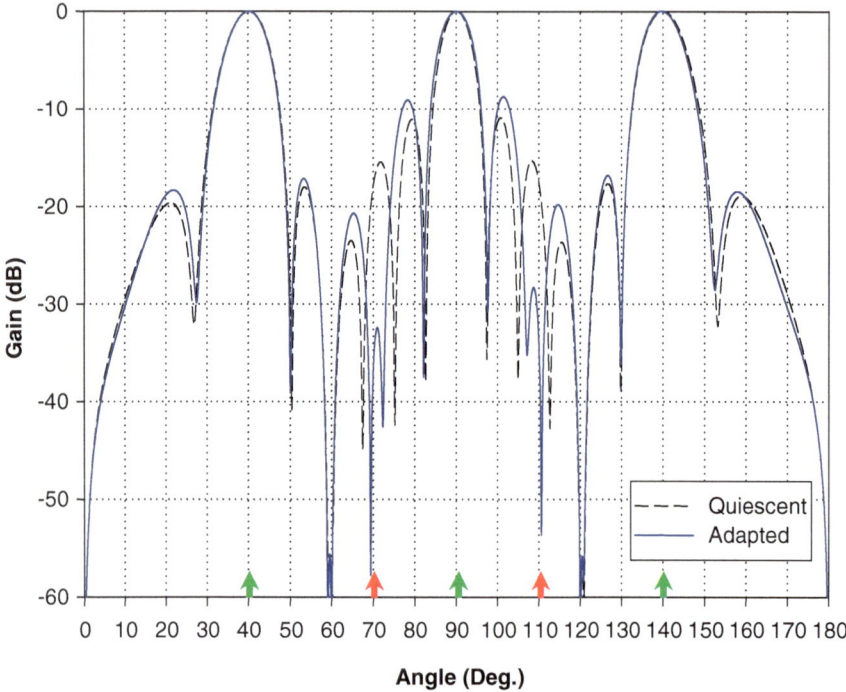

Fig. 34 Adapted beam pattern of 16-element linear dipole array. Three desired signals (40°, 90°, 140°; 1) and two probing sources (70°, 110°; 100)

pattern is shown in Fig. 34. Again, each of two probing sources is suppressed without any distortion in each of the three main lobes towards the desired signals.

Figure 35 shows the adapted pattern for three desired signals (40°, 90°, 140°; 1 each) and four probing sources (70°, 80°, 100°, 110°; 1000, 1000, 800, 500). It can be observed that even in this complex scenario, the mainlobes towards three desired signals are maintained with accurate nulls towards each of the probing sources.

This demonstrates the capability of the modified improved LMS algorithm in generating correct adapted pattern even when the mutual coupling is taken into account. This helps phased array in catering complicated signal environments consisting of multiple sources.

5.2 *Parallel-in-Echelon Configuration*

In this sub-section, the parallel-in-echelon dipole configuration is considered. Accordingly the self and mutual impedance are calculated before the generation of adapted pattern. Figure 36 shows the output noise power of 16-element linear

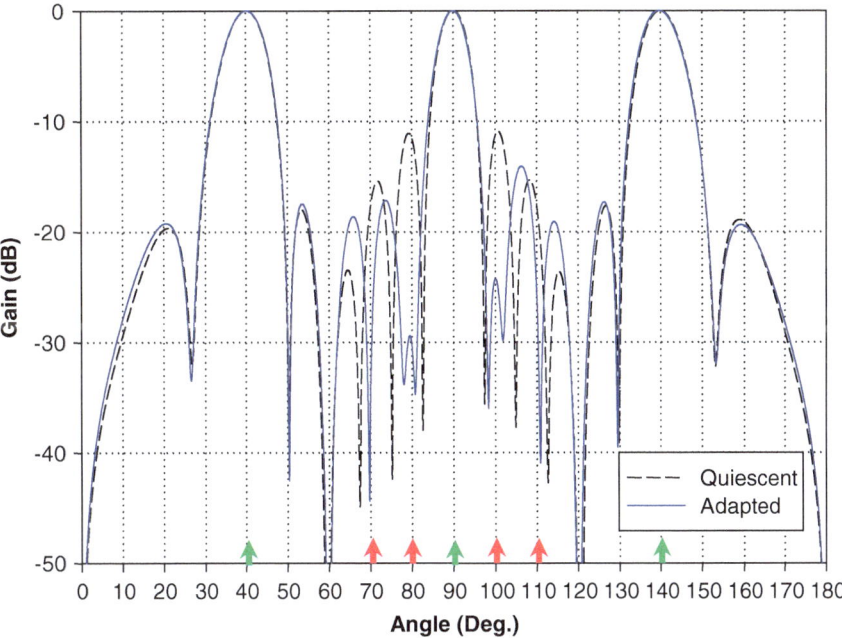

Fig. 35 Adapted beam pattern of 16-element linear dipole array. Three desired signals (40°, 90°, 140°; 1 each) and four probing sources (70°, 80°, 100°, 110°; 1000, 1000, 800, 500)

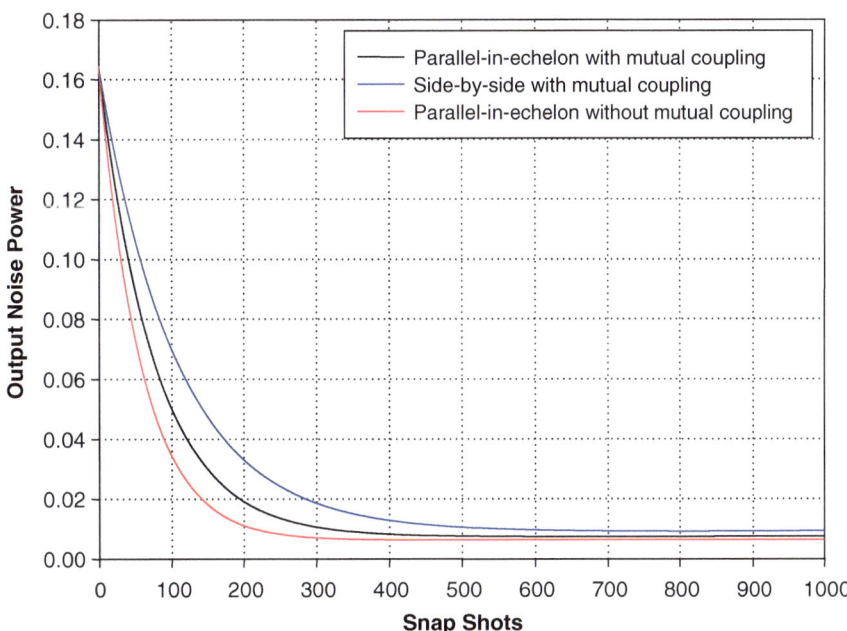

Fig. 36 Output noise power of 16-element linear dipole array. One desired signal (90°; 1) and two probing sources (35°, 145°; 100 each)

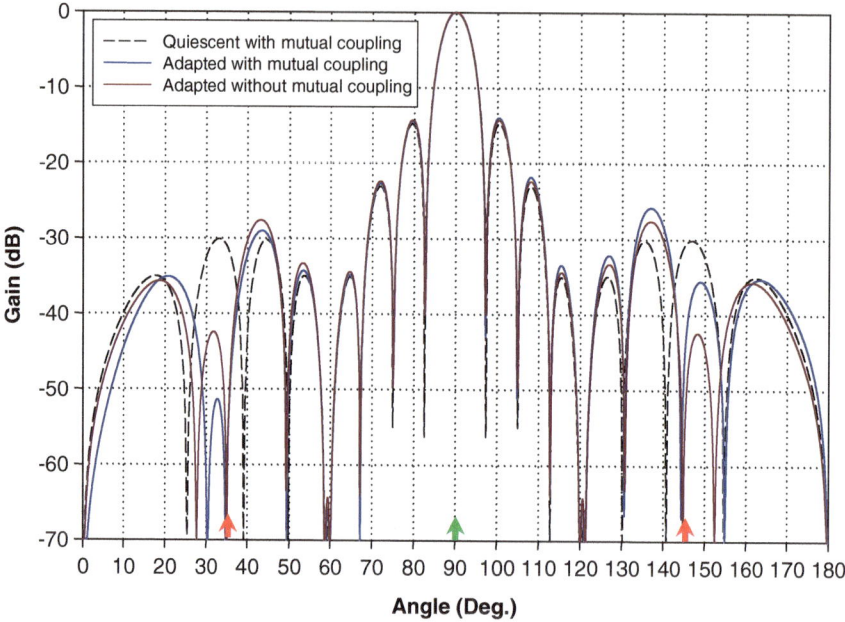

Fig. 37 Adapted beam pattern of 16-element linear parallel-in-echelon dipole array. One desired signal (90°; 1) and two probing sources (35°, 145°; 100)

dipole array with one desired signal (90°; 1) and two probing sources (35°, 145°; 100 each). The output noise power of parallel-in-echelon is compared with that of side-by-side configuration. As expected the output noise power of parallel-in-echelon is lower than that of side-by-side configuration. Moreover, the performance of parallel-in-echelon array is compared for with and without mutual coupling. It can be observed that the performance of parallel-in-echelon array without mutual coupling effect is best as compared to the case of parallel-in-echelon array with mutual coupling and side-by-side array with mutual coupling.

The corresponding adapted pattern is shown in Fig. 37. The adapted pattern consists of deep nulls at both the probing source directions without any distortion in the main lobe.

The performance of 16-element linear parallel-in-echelon dipole array is analyzed in terms of output SINR (Fig. 38). The signal scenario consists of one desired signal (90°; 1) and two probing sources (45°, 135°; 100 each). The result of parallel-in-echelon is compared with the side-by-side configuration with mutual coupling. As expected, the output SINR in parallel-in-echelon is better than that of side-by-side configuration (with mutual coupling). Further the performance of parallel-in-echelon without mutual coupling is better than the case with mutual coupling. The corresponding adapted pattern for the same signal scenario is shown in Fig. 39. It can be observed that the mainlobe is maintained with deep nulls placed towards each of the two probing sources.

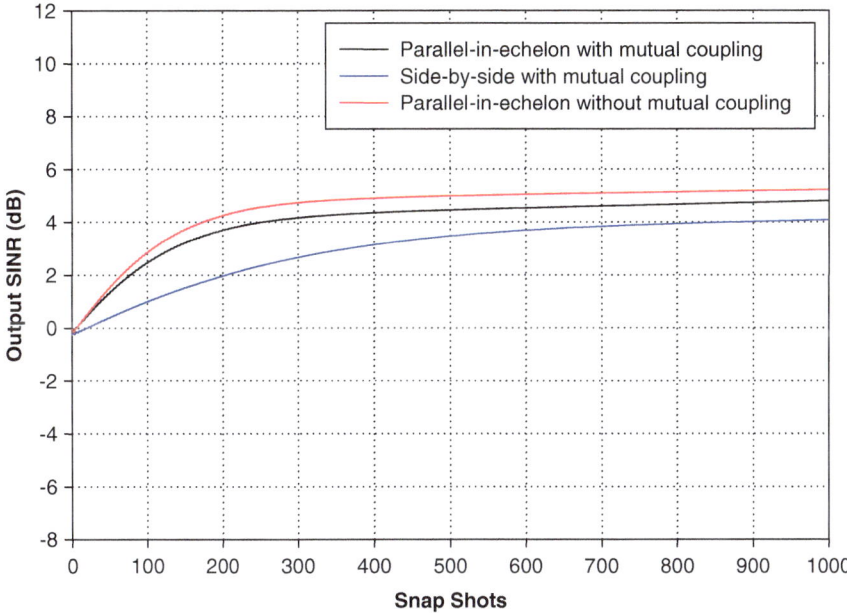

Fig. 38 Output SINR of 16-element linear dipole array. One desired signal (90°; 1) and two probing sources (45°, 135°; 100)

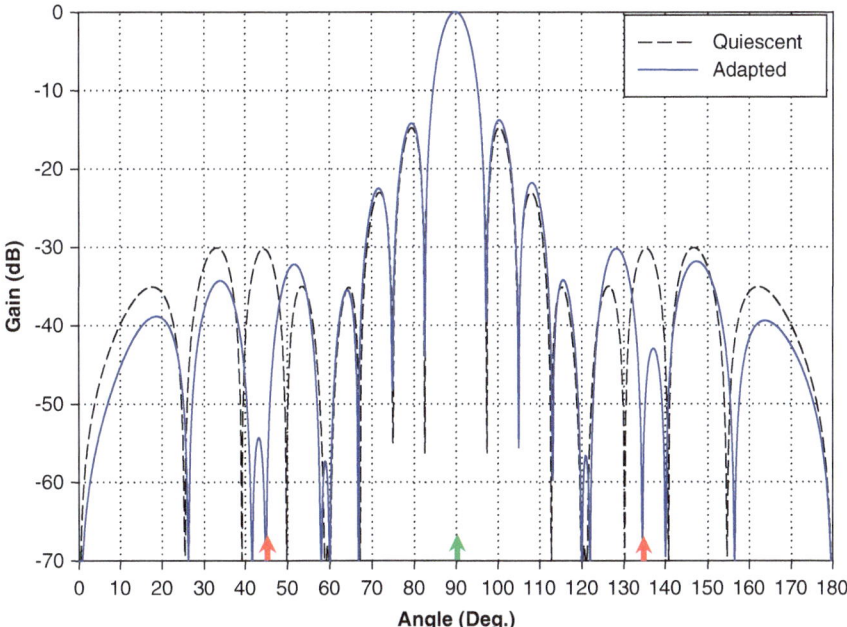

Fig. 39 Adapted beam pattern of 16-element linear parallel-in-echelon dipole array. One desired signal (90°; 1) and two probing sources (45°, 135°; 100)

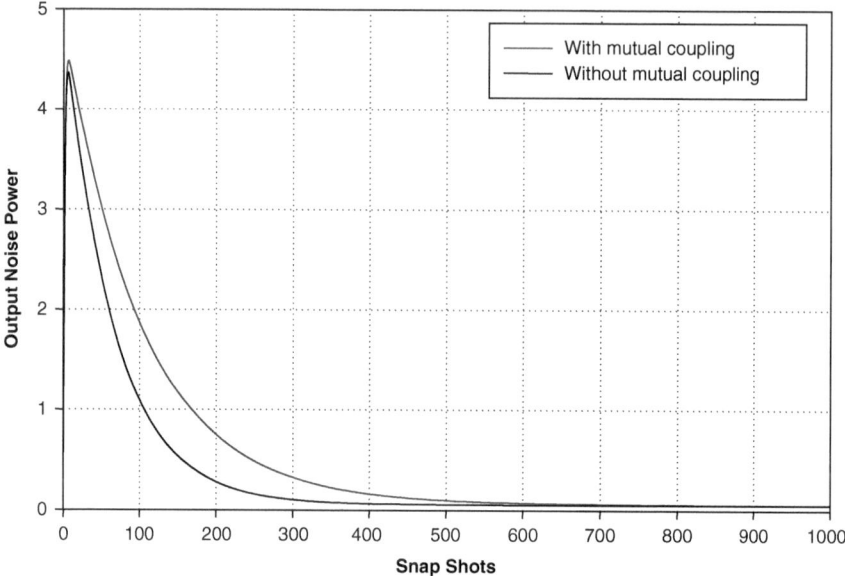

Fig. 40 Output noise power of 16-element linear parallel-in-echelon dipole array. Two desired signals (40°, 140°; 1 each) and one probing source (95°; 100)

Next the signal scenario consists of two desired signals and one probing source. Figure 40 shows output noise power of 16-element linear parallel-in-echelon dipole array with two desired signals (40°, 140°; 1 each) and one probing source (95°; 100). The corresponding adapted pattern is shown in Fig. 41.

The adapted pattern is as per the expectations. Next a similar signal scenario but with different impinging angles is considered. The two desired signals (40°, 90°; 1 each) and one probing source (110°; 100) are assumed to be incident on 16-element linear parallel-in-echelon dipole array.

Figure 42 shows the resultant output SINR for with and without mutual coupling. The performance of array is much better for without mutual coupling case. The corresponding adapted pattern is shown in Fig. 43. It is apparent that the array maintains the gain towards desired sources and at the same time null is placed towards the probing source without any distortion in the mainlobes.

The number of probing sources are increased to three (45°, 80°, 110°; 1000, 600, 800) keeping the two desired signals (90°, 140°; 1 each). The adapted pattern of 16-element linear parallel-in-echelon dipole array is shown in Fig. 44. The pattern is as per the expectations.

Next the number of desired signals is increased to three (40°, 90°, 140°; 1 each) and keeping two probing sources (70°, 100°; 1000 each). The resultant adapted pattern is shown in Fig. 45. Again, each of two probing sources is suppressed without any distortion in the three main lobes towards each of the desired signals. The overall pattern remains undisturbed.

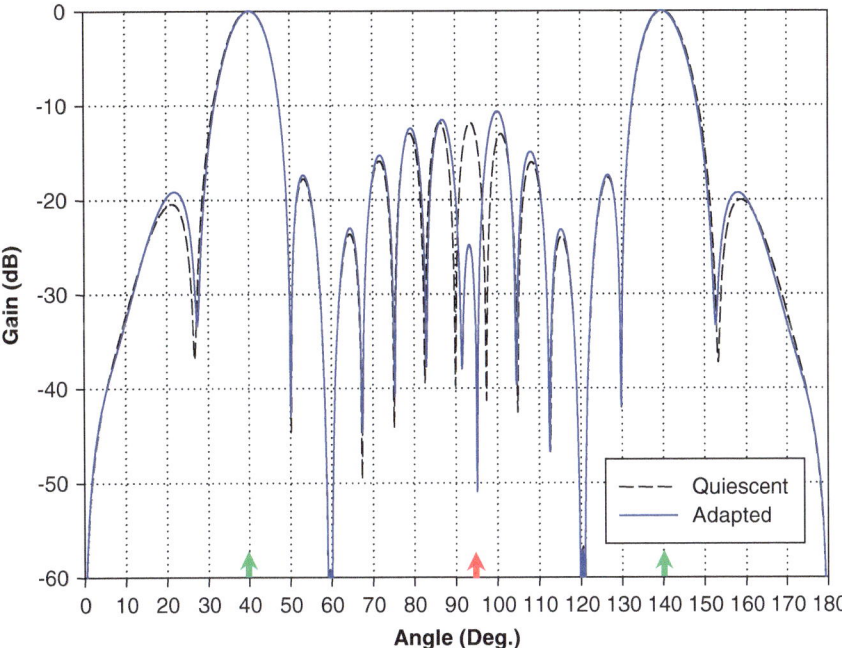

Fig. 41 Adapted beam pattern of 16-element linear parallel-in-echelon dipole array. Two desired signals (40°, 140°; 1 each) and one probing source (95°; 100)

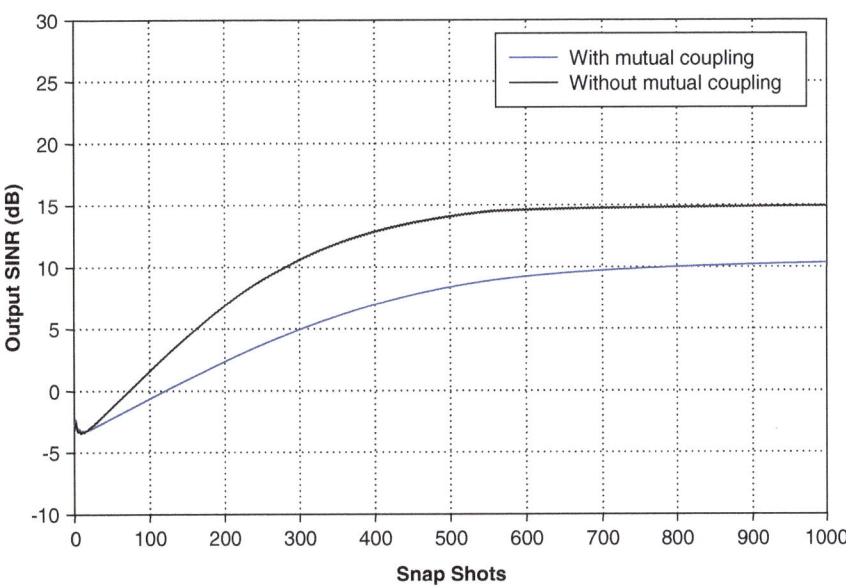

Fig. 42 Output SINR of 16-element linear parallel-in-echelon dipole array. Two desired signals (40°, 90°; 1 each) and one probing source (110°; 100)

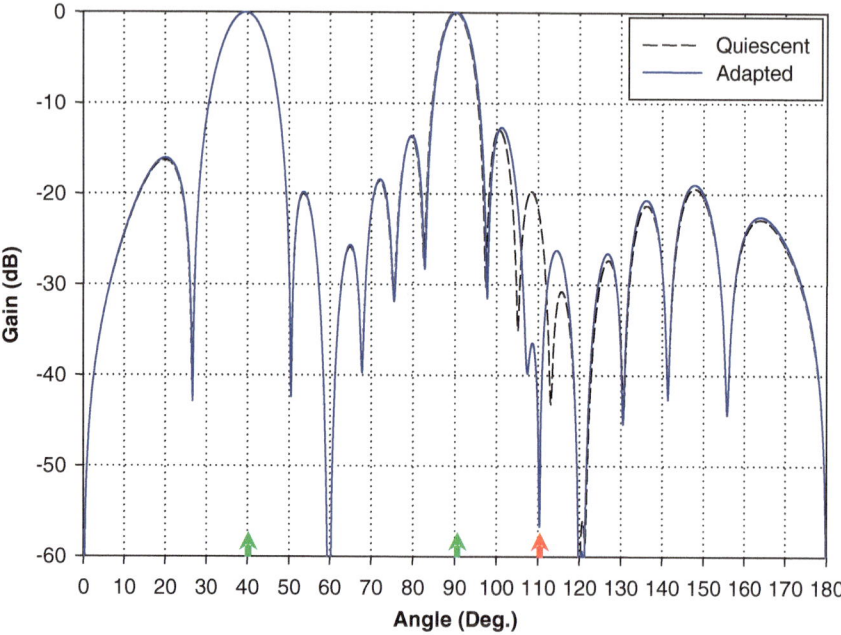

Fig. 43 Adapted beam pattern of 16-element linear parallel-in-echelon dipole array. Two desired signals (40°, 90°; 1 each) and one probing source (110°; 100)

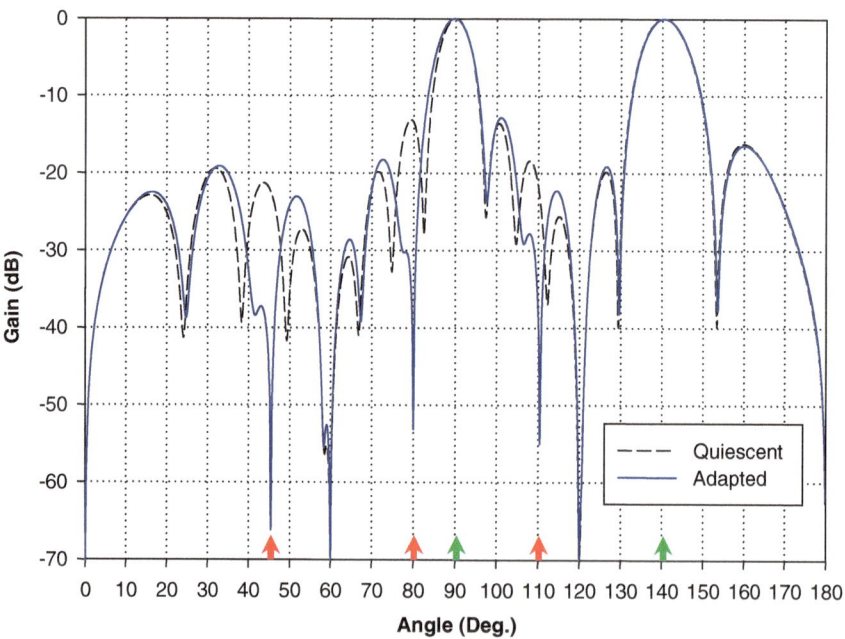

Fig. 44 Adapted beam pattern of 16-element linear parallel-in-echelon dipole array. Two desired signals (90°, 140°; 1 each) and three probing sources (45°, 80°, 110°; 1000, 600, 800)

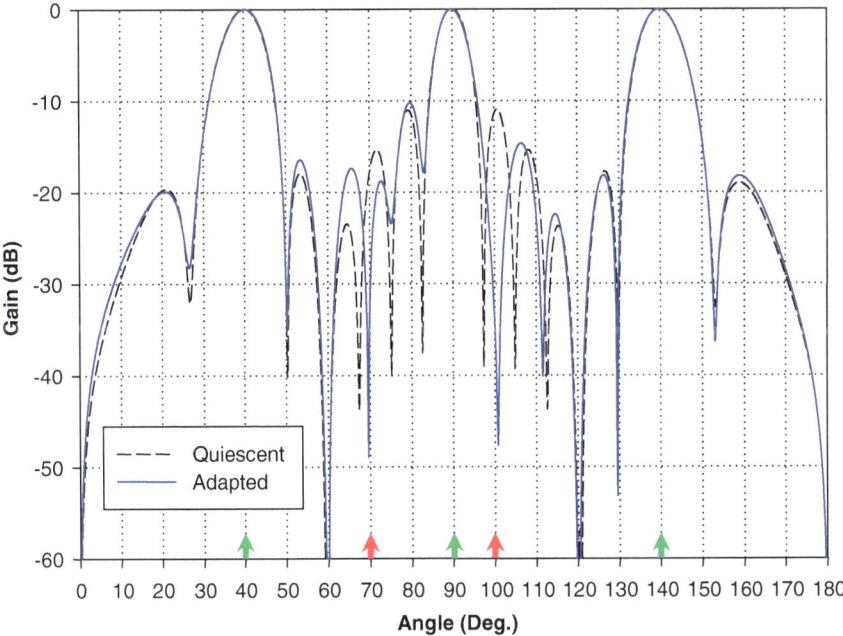

Fig. 45 Adapted pattern of 16-element linear parallel-in-echelon dipole array. Three desired signals (40°, 90°, 140°; 1 each) and two probing sources (70°, 100°; 1000 each)

Figure 46 shows the adapted pattern of three desired signals (40°, 90°, 140°; 1 each) and three probing sources (70°, 100°, 110°; 1000 each). It is apparent that the adapted pattern is as per the expectations. Next a more complicated signal scenario is considered. Figure 47 shows the adapted pattern of three desired signals (40°, 90°, 140°; 1 each) and four probing sources (70°, 80°, 100°, 110°; 1000 each). It can be observed that even three desired signals are maintained with accurate nulls towards each of the four probing sources.

6 Edge Effect in Array Processing

The edge effect in finite antenna arrays has been the subject of interest towards performance optimization (Hansen 1996, 1999). For a finite array, the behavior of edge elements is in general different from the rest of the array. The real part of the input impedance of antenna array does not depend on the element position in the array while this is not true for the imaginary part of the impedance matrix. This effect alters the array response towards the impinging signals (Hansen 2004).

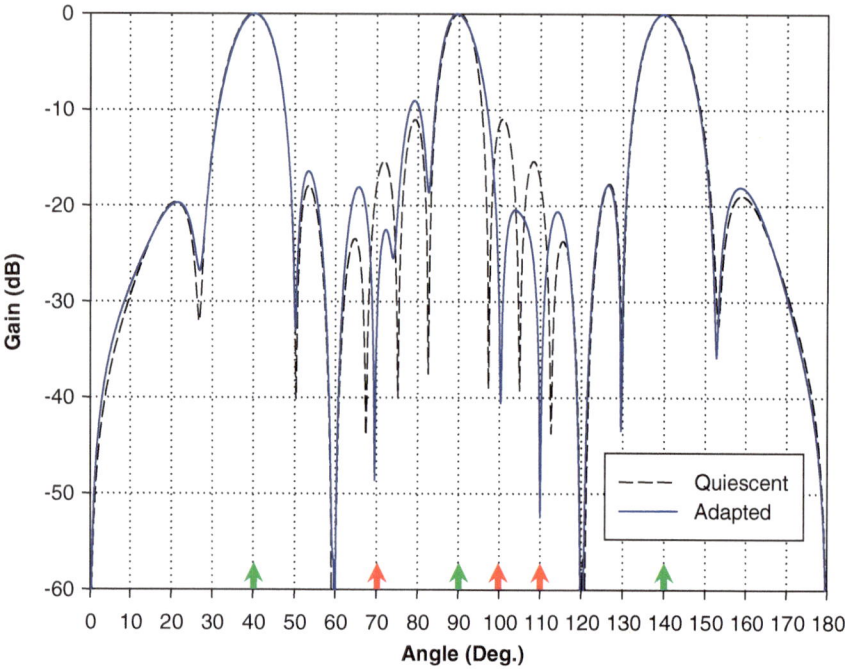

Fig. 46 Adapted pattern of 16-element linear parallel-in-echelon dipole array. Three desired signals (40°, 90°, 140°; 1) and three probing sources (70°, 100°, 110°; 1000)

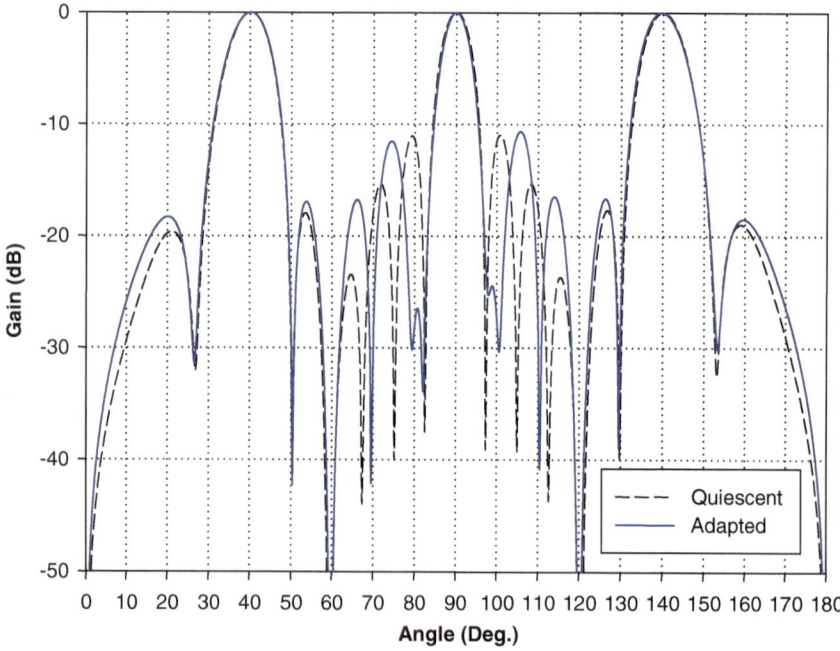

Fig. 47 Adapted beam pattern of 16-element linear parallel-in-echelon dipole array. Three desired signals (40°, 90°, 140°; 1 each) and four probing sources (70°, 80°, 100°, 110°; 1000 each)

Including edge effects, the array response towards the signal incident at an angle θ is given by

$$S_t = f(\theta,\, \phi) \cdot \left(C_{ii} \cdot e^{-j(kd\cos\theta + \alpha)} + \sum_{k=1;\, k \neq i}^{N} S_{ik} \cdot e^{-j(kd\cos\theta + \alpha)} \right) \tag{32}$$

where, S_{ik} is the coupling scattering coefficient (Rabinovich and Alexandrov 2012) is given by

$$S_{ik} = \begin{cases} 0 & i = k \\ \frac{2 \cdot Z_{ik}}{(Z_{ii}+1) \cdot (Z_{kk}+1) - Z_{ik} \cdot Z_{ki}} & \text{otherwise} \end{cases}$$

k is the wave number, d is the inter-element spacing, Z is the impedance matrix consisting of both self and mutual impedances, C_{ii} denotes the coupling of the array aperture to the output transmission line, α is the phase excitation, $f(\theta,\, \phi)$ is the radiation pattern of the antenna element.

6.1 Dipole Array in Side-by-Side Configuration

In this sub-section, a 16-element half-wavelength dipole array with side-by-side geometric configuration is considered. The inter-element spacing is 0.484λ. The edge and mutual coupling effects are included in computations for the output noise power, output SINR, and adapted pattern for different signal environments.

Figure 48 shows the adapted and quiescent patterns for signal environment consisting of one desired source (90°, 1) and one probing source (110°; 100). The array is capable of generating desired adapted pattern even when edge effect and mutual coupling between antenna elements are taken into account.

Figure 49 compares the output noise power of the dipole array with and without edge effect. The signal scenario consists of one desired signal (90°; 1) and two probing sources (45°; 135°, 150, 200). It is apparent that the output noise power is little higher with edge effects. The corresponding output SINR is shown in Fig. 50. As expected output SINR degrades slightly when edge effect is taken into account. However the variation is not much significant.

The adapted and quiescent patterns of the array is shown in Fig. 51. The mainlobe is maintained towards the desired source with accurate null placement towards each of the probing sources.

A signal scenario consisting of two desired signals (40°, 140°; 1) and one probing source (80°; 100) is considered next. The adapted pattern is generated including edge and mutual coupling effect. From Fig. 52, it can be observed that the array places accurate null towards the probong signal, maintaining mainlobes towards each of the desired signals.

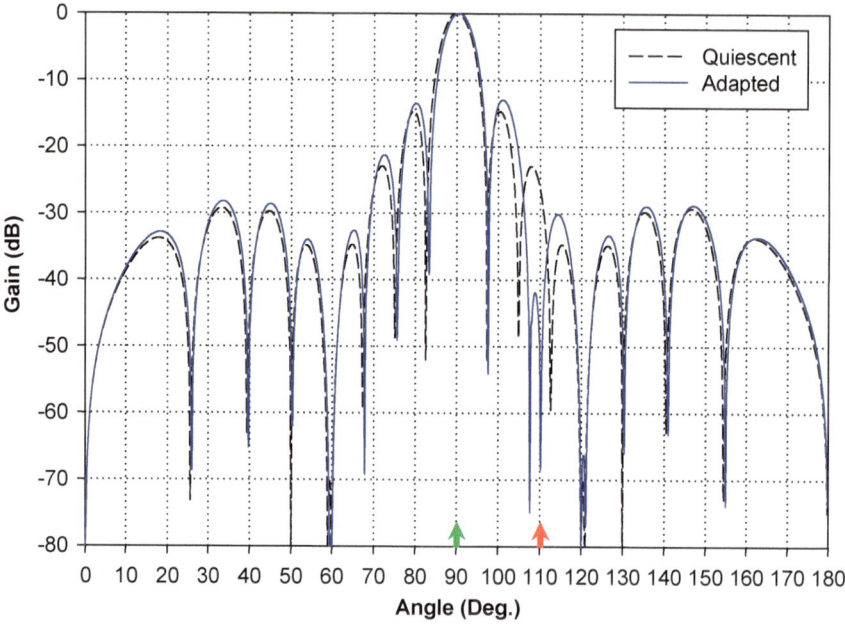

Fig. 48 Adapted pattern of 16-element linear side-by-side half-wavelength dipole array including edge and mutual coupling effects. One desired signal (90°, 1) and one probing source (110°; 100)

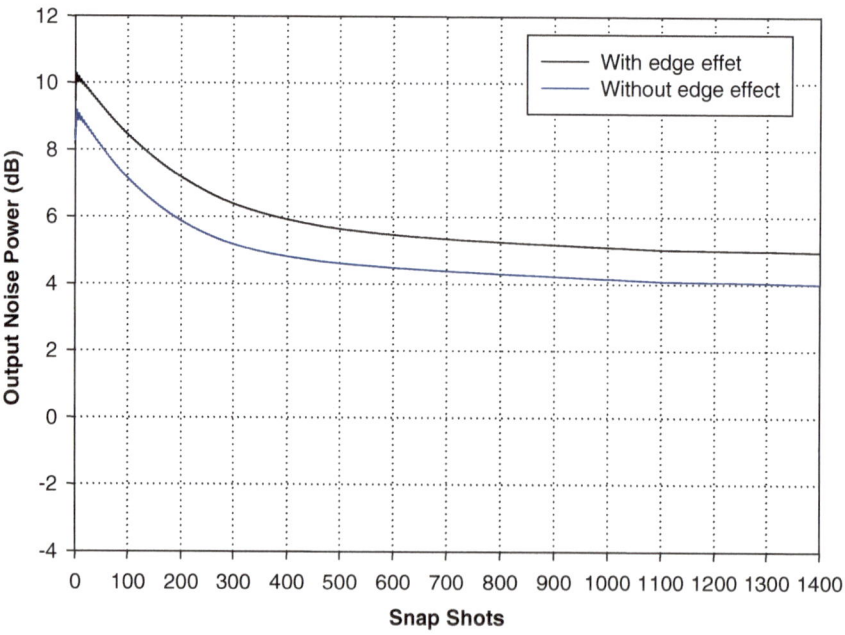

Fig. 49 Output noise power of 16-element linear side-by-side dipole array including edge and mutual coupling effects. One desired signal (90°; 1) and two probing sources (45°, 135°; 150, 200)

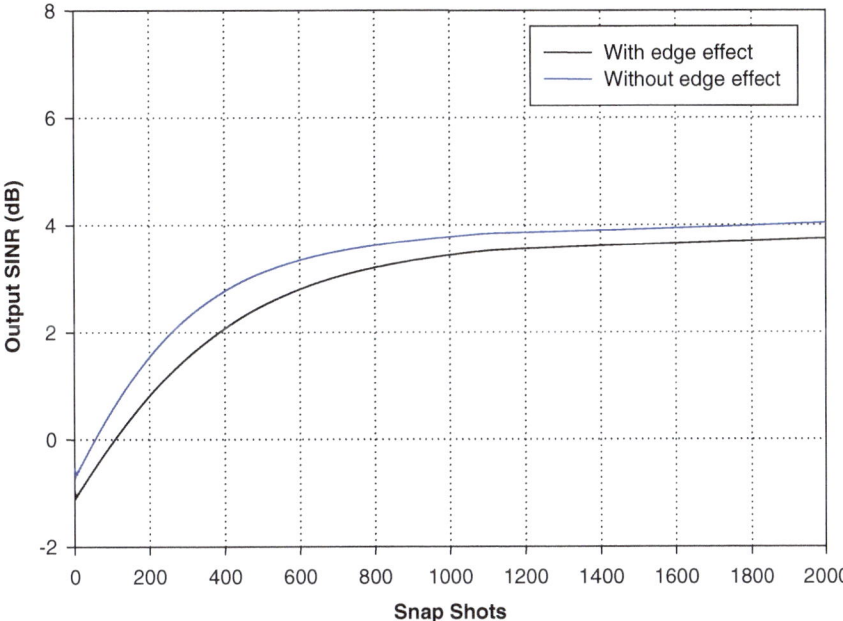

Fig. 50 Output SINR of 16-element linear side-by-side dipole array including edge and mutual coupling effects. One desired signals (90°; 1) and two probing sources (45°, 135°; 150, 200)

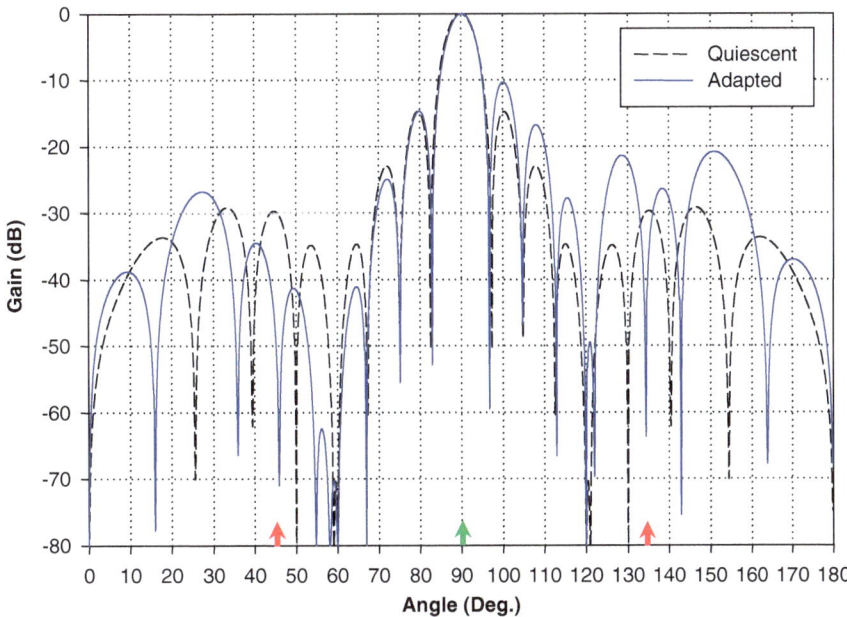

Fig. 51 Adapted pattern of 16-element linear side-by-side dipole array including edge and mutual coupling effects. One desired signal (90°; 1) and two probing sources (45°, 135°; 150, 200)

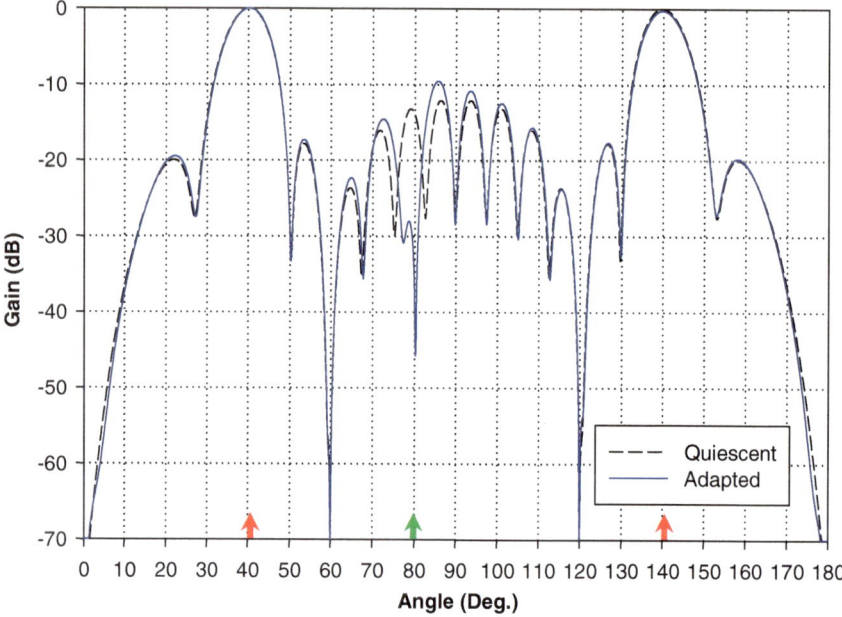

Fig. 52 Adapted pattern of 16-element linear side-by-side dipole array including edge and mutual coupling effects. Two desired signals (40°, 140°; 1) and one probing source (80°; 100)

Next, number of probing signals are increased to two. Figure 53 shows the adapted pattern of 16-element linear side-by-side dipole array including edge and mutual coupling effects. The signal environment consists of two desired signals (35°, 145°; 1 each) and two probing sources (75°, 90°; 800, 1000). The adapted pattern shows mainlobe towards each of the desired signals with accurate and deep nulls towards each of probing sources. These probing sources will not be able to receive any signal from the array. In other words the array is invisible to them.

Figure 54 presents the adapted pattern for two desired signals (40°, 140°; 1 each) and three probing sources (70°, 95°, 115°; 600, 800, 700) impinging the array. It is apparent that array is able to cater to the signal scenario even when edge and mutual coupling effects are taken into account. This demonstrates the efficiency of adaptive algorithm used for weight adaptation in array.

Next three desired signals (40°, 90°, 140°; 1 each) and one probing source (110°; 100) is considered. Figure 55 shows the adapted pattern of 16-element linear side-by-side dipole array including edge and mutual coupling effects. It may be observed that the array pattern consisting of three mainlobes towards each of desired sources while the probing source is actively cancelled out by placing a deep null towards it.

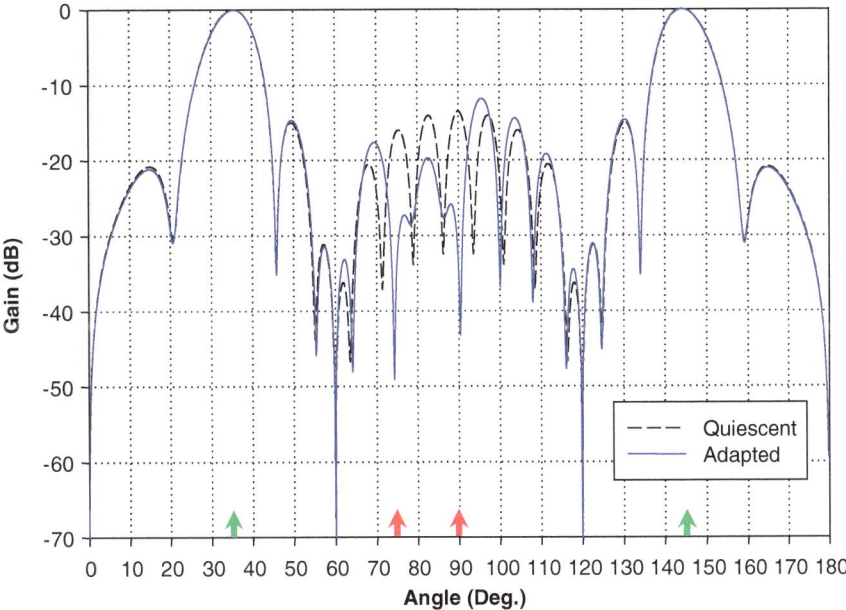

Fig. 53 Adapted pattern of 16-element linear side-by-side dipole array including edge and mutual coupling effects. Two desired signals (35°, 145°; 1 each) and two probing sources (75°, 90°; 800, 1000)

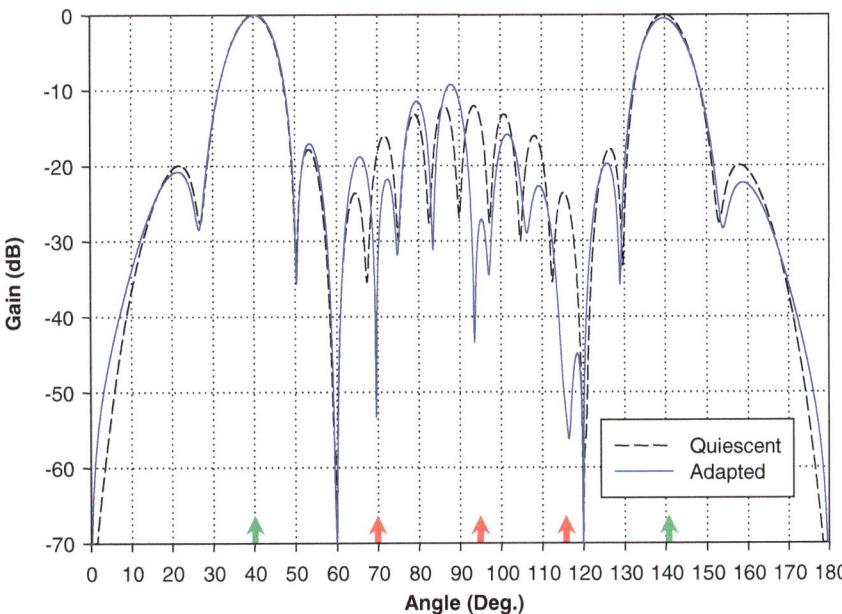

Fig. 54 Adapted pattern of 16-element linear side-by-side dipole array including edge and mutual coupling effects. Two desired signals (40°, 140°; 1 each) and three probing sources (70°, 95°, 115°; 600 800, 700)

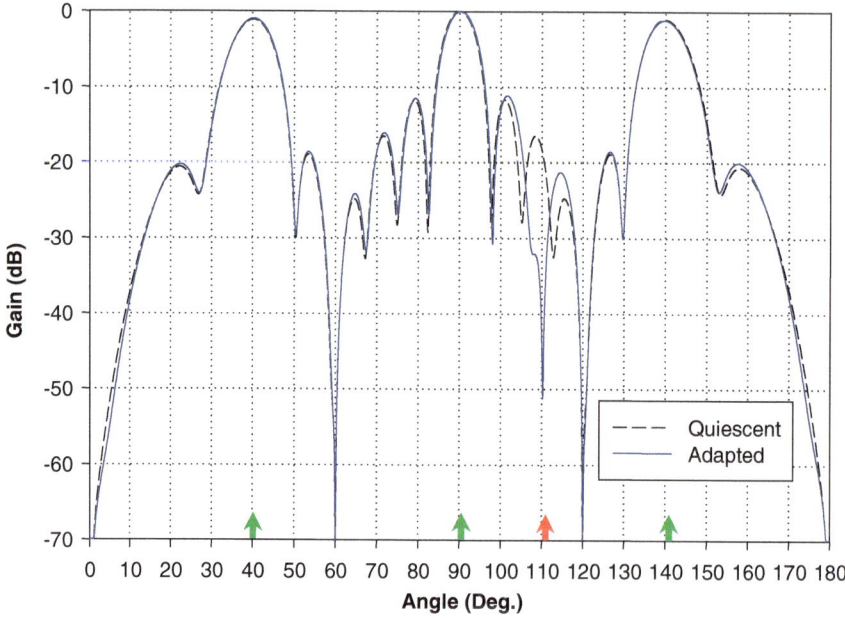

Fig. 55 Adapted pattern of 16-element linear side-by-side dipole array including edge and mutual coupling effects. Three desired signals (40°, 90°, 140°; 1 each) and one probing source (110°; 100)

6.2 Dipole Array in Parallel-in-Echelon Configuration

In this section, the configuration of dipole array is taken as parallel-in-echelon. The inter-element spacing is 0.484λ.

The height h is taken as 0.25λ from reference plane. The performance of 16-element half-wavelength dipole array is analyzed including the edge and mutual coupling effects. Figure 56 shows the output noise power of the array for signal environment consisting of one desired signal (90°; 1) and one probing source (35°; 100). As in Sect. 6.1, the output noise power is little higher as compared to the case when edge effects are ignored. The corresponding output SINR is shown in Fig. 57. As expected the output SINR of array reduces a little when edge effects are taken into account. However degradation is by 0.1 dB. The corresponding adapted pattern of 16-element linear parallel-in-echelon dipole array is shown in Fig. 58.

Next, a signal environment consisting of one desired signal (90°; 1) and two probing sources (45°, 135°; 100 each) is considered. Figure 59 shows the adapted pattern along with the quiescent pattern. It is apparent that the array is able to cater the signal scenario, by actively cancelling both the probing sources with main lobe towards the desired direction.

Figure 60 presents the adapted pattern when there are two desired signals (40°, 140°; 1 each) and one probing source (80°; 200). It may be observed that array

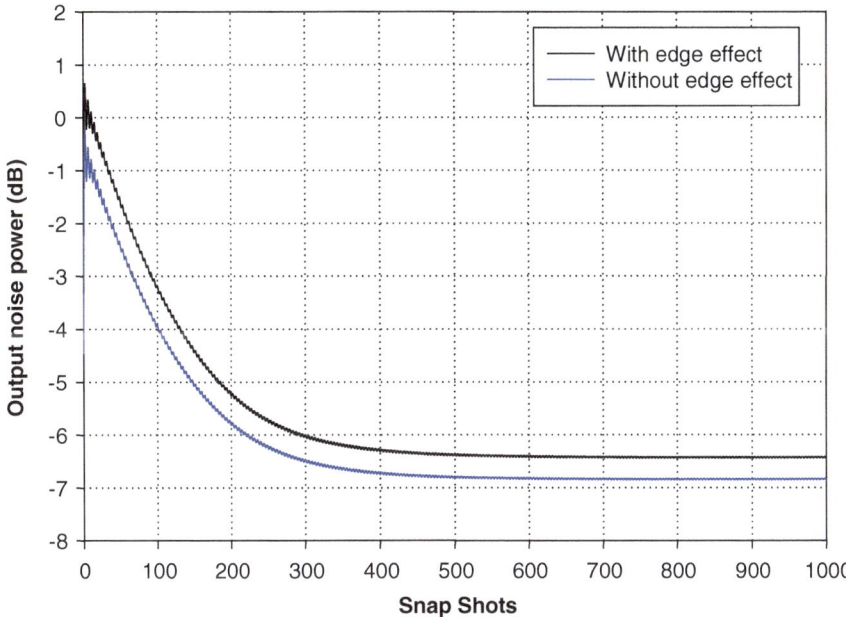

Fig. 56 Output noise power of 16-element linear parallel-in-echelon dipole array including edge and mutual coupling effects. One desired signal (90°; 1) and one probing source (35°; 100)

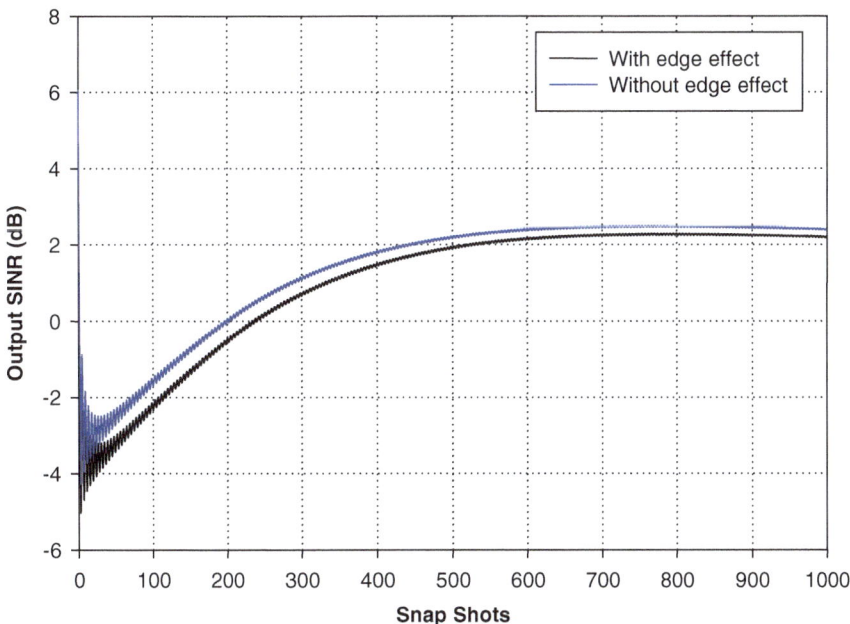

Fig. 57 Output SINR of 16-element linear parallel-in-echelon dipole array including edge and mutual coupling effects. One desired signal (90°; 1) and one probing source (35°; 100)

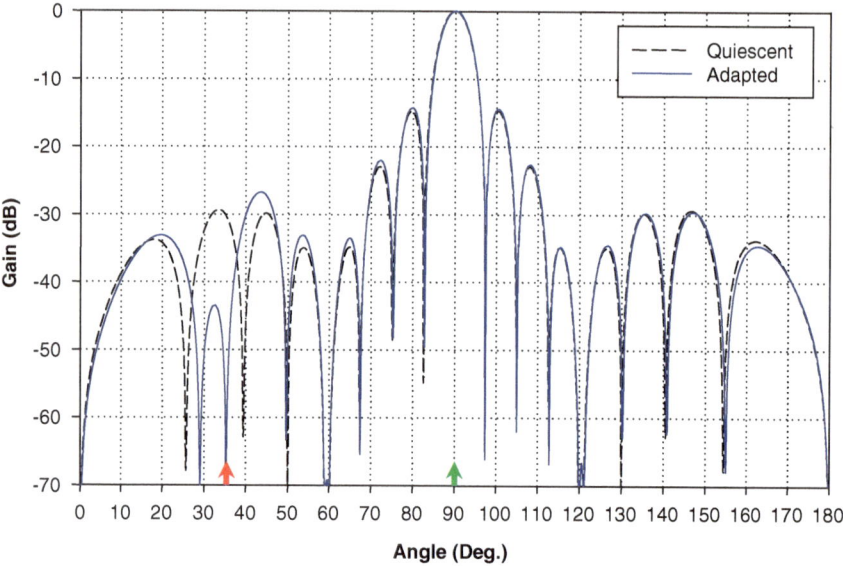

Fig. 58 Adapted pattern of 16-element linear parallel-in-echelon dipole array including edge and mutual coupling effects. One desired signal (90°; 1) and one probing source (35°; 100)

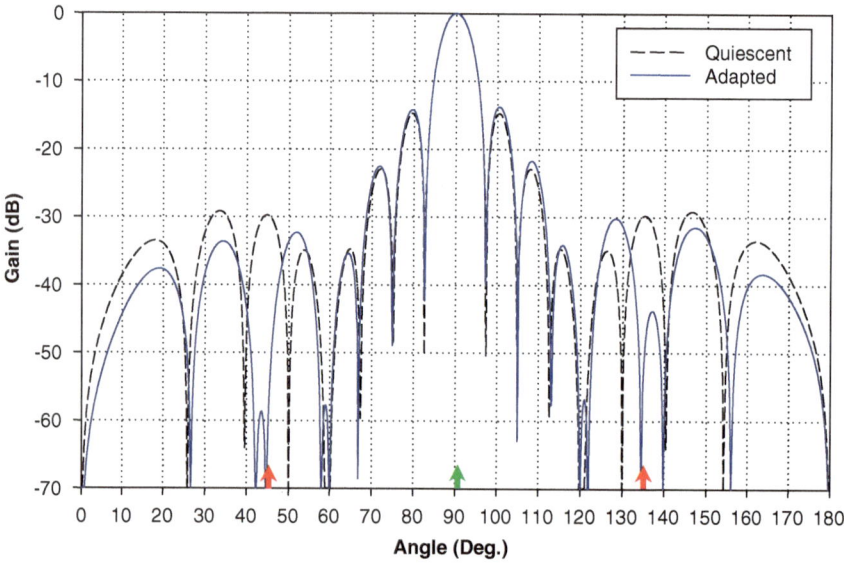

Fig. 59 Adapted pattern of 16-element linear parallel-in-echelon dipole array including edge and mutual coupling effects. One desired signal (90°; 1) and two probing sources (45°, 135°; 100 each)

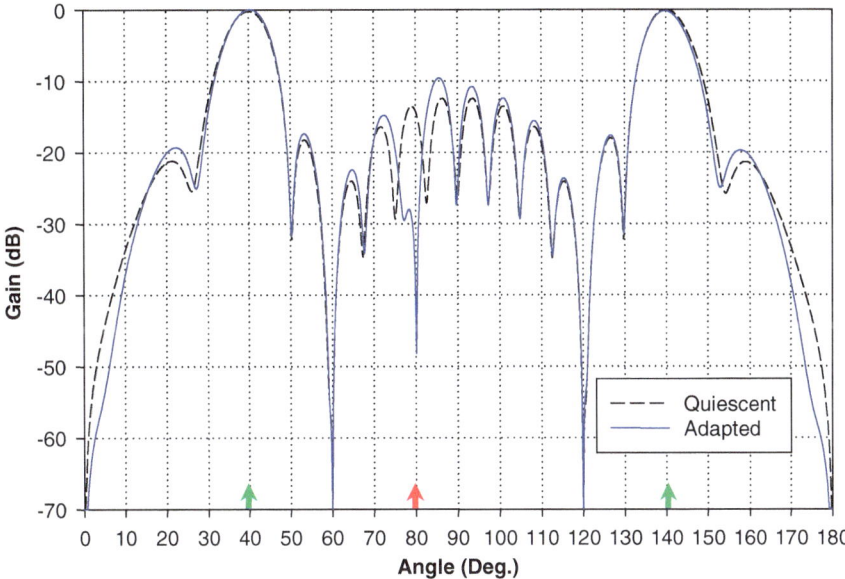

Fig. 60 Adapted pattern of 16-element linear parallel-in-echelon dipole array including edge and mutual coupling effects. Two desired signals (40°, 140°; 1 each) and one probing source (80°; 200)

maintains mainlobe towards each of the desired sources with a deep null towards the probing source. Figure 61 presents the adapted pattern when two desired signals (35°, 145°; 1 each) and two probing sources (75°, 105°; 100, 200) impinge the array.

Next a more complicated signal environment consisting of three desired signals (40°, 90°, 140°; 1 each) and one probing source (110°; 100) is considered. Figure 62 shows the adapted pattern that demonstrates the efficiency of array in catering such signal scenario. The array maintains mainlobe towards each of the desired sources, with −30 dB null in the probing direction. Next number of probing signals is increased to two with three desired signals. Figure 63 presents the adapted pattern of 16-element linear parallel-in-echelon dipole array including edge and mutual coupling effects. The signal scenario considered is three desired signals (40°, 90°, 140°; 1 each) and two probing sources (70°, 110°; 150, 200).

It may be observed that array efficiently nullify both the probing signals and simultaneously maintains mainlobe towards each of the desired source. Thus one can infer that the modified improved LMS algorithm is an efficient weight adaptation algorithm, which can efficiently cater to arbitrary signal environment including edge and mutual coupling effect.

This capability of adaptive algorithm can be used towards active radar cross section (RCS) reduction of platforms over which phased arrays are mounted.

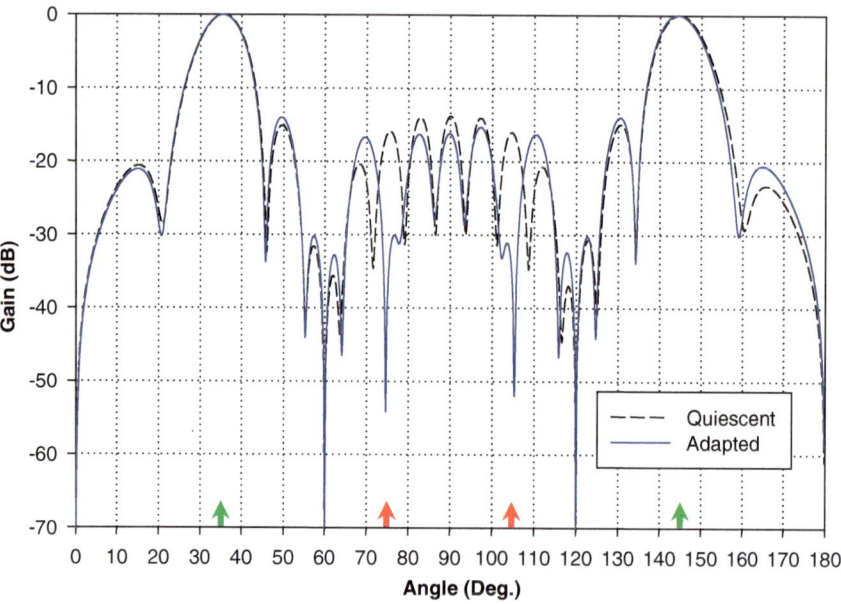

Fig. 61 Adapted pattern of 16-element linear parallel-in-echelon dipole array including edge and mutual coupling effects. Two desired signals (35°, 145°; 1 each) and two probing sources (75°, 105°; 100, 200)

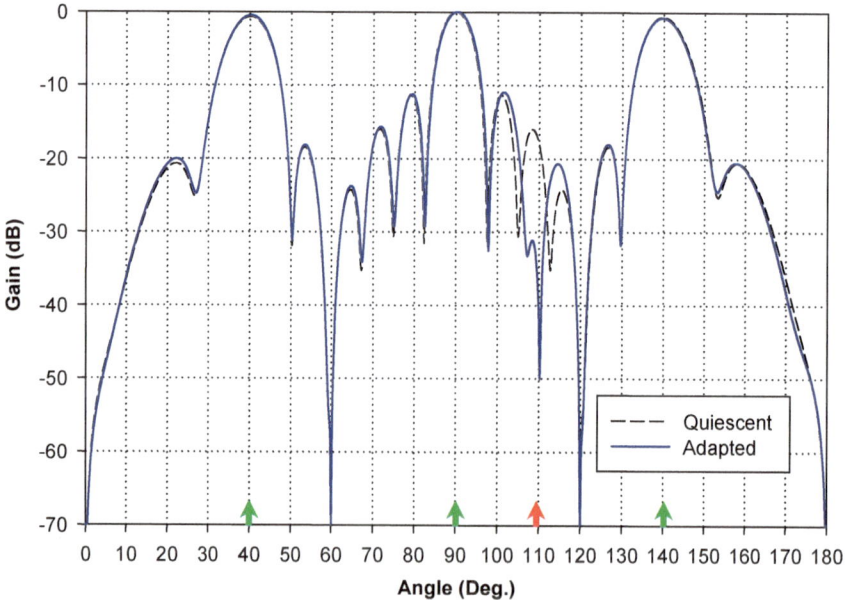

Fig. 62 Adapted pattern of 16-element linear parallel-in-echelon dipole array including edge and mutual coupling effects. Three desired signals (40°, 90°, 140°; 1 each) and one probing source (110°; 100)

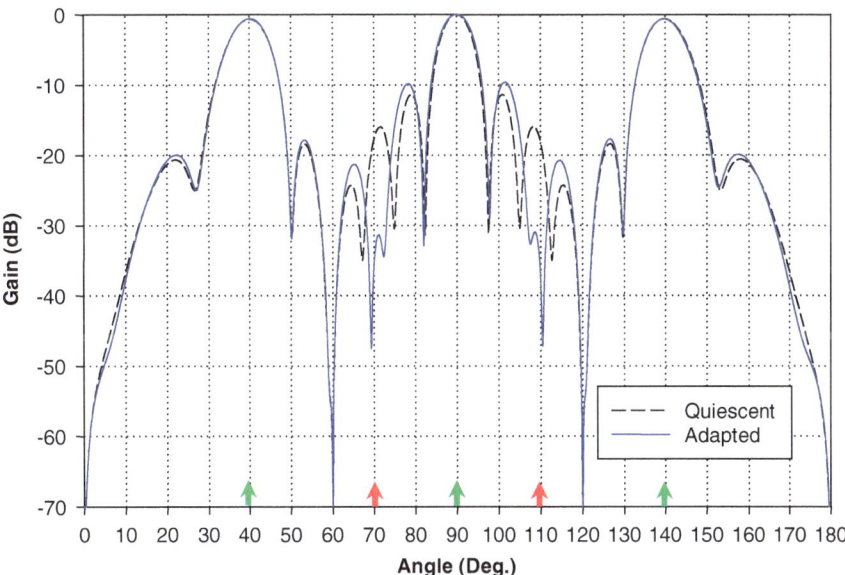

Fig. 63 Adapted pattern of 16-element linear parallel-in-echelon dipole array including edge and mutual coupling effects. Three desired signals (40°, 90°, 140°; 1 each) and two probing sources (70°, 110°; 150, 200)

7   Conclusion

This brief demonstrate the efficiency of modified improved LMS algorithm in uniform linear array of dipoles and monopoles for active cancellation of multiple probing sources, and simultaneously maintaining gain towards each of the desired sources. The weight adaptation and hence the adapted pattern generation depends on the design configuration of array, signal scenario, and step-size of iterative calculation. The array with an efficient modified improved LMS algorithm is shown to be capable of catering arbitrary narrowband signal environments. The performance of array does not degrade even when the probing source impinges from the angle that lies within the mainlobe of the quiescent pattern. The nulls are placed accurately towards each of probing directions without any distortion in the mainlobe and sidelobe levels.

The role of edge elements and coupling between the elements is taken into account. Both edge effect and mutual coupling between the array elements alters the steering vector and hence the weight adaptation. This affects the suppression capability of the phased array. It is shown here that the adaptive nulling performance is well maintained for an arbitrary signal environment and even in the presence of mutual coupling effect. This demonstrates the capability of the algorithm in actively reducing the detectability of phased array.

References

Adve, R.S., and T.K. Sarkar. 2000. Compensation for the effects of mutual coupling on direct data domain adaptive algorithms. *IEEE Transactions on Antennas and Propagation* 48: 86–94.

Balanis, C.A. 2005. *Antenna theory, analysis and design*, 3rd edn, 1117 p. Hoboken, New Jersey: Wiley. ISBN:0-471-66782-X.

Bernardi, G., M. Felaco, M.D. Urso, L. Thimmoneri, A. Ferina, and E.F. Meliado. 2011. A simple strategy to tackle mutual coupling and platform effects in surveillance systems. *Progress In Electromagnetics Research C* 20: 1–15.

Compton, Jr., R.T. 1982. A method of choosing element patterns in an adaptive array. *IEEE Transactions on Antennas and Propagation* AP-30: 489–493.

Elliott, R.S., and G.J. Stern. 1981a. The design of microstrip dipole arrays including mutual coupling, Part I: Theory. *IEEE Transactions on Antennas and Propagation* AP-29: 757–760.

Elliott, R.S., and G.J. Stern. 1981b. The design of microstrip dipole arrays including mutual coupling, Part II: Experiment. *IEEE Transactions on Antennas and Propagation* AP-29: 761–765.

Ganz, M.W., R.L. Moses, and S.L. Wilson. 1990. Convergence of the SMI and the diagonally loaded SMI algorithms with weak interference. *IEEE Transactions on Antennas and Propagation* 38: 394–399.

Godara, L.C. 2004. *Smart antennas*, 448 p. Washington DC: CRC Press. ISBN:0-8493-1206-X.

Griffith, K.A., and I.J. Gupta. 2008. Effect of mutual coupling on the performance of GPS AJ antennas. in *Proceedings of IEEE Symposium on Position, Location, and Navigation Symposium*, Monterey, CA, May, pp. 871–877.

Gupta, I.J., and A.A. Ksienski. 1982. Dependence of adaptive array performance on conventional array design. *IEEE Transactions on Antennas and Propagation* AP-30: 549–553.

Gupta, I.J., J.A. Ulrey, and E.H. Newman. 2005. Effects of antenna element bandwidth on adaptive array performance. *IEEE Transactions on Antennas and Propagation* AP-53: 2332–2336.

Hansen, R.C. 1996. Finite array scan impedance Gibbsian models. *Radio Science* 31: 1631–1637.

Hansen, R.C. 1999. Anomalous edge effects in finite arrays. *IEEE Transactions on Antennas and Propagation* 47(3): 549–554.

Hansen, R.C. 2004. Linear connected arrays. *IEEE Antenna and Wireless Propagation Letters* 3: 154–156.

Hui, H.T. 2004. A practical approach to compensate for the mutual coupling effect of an adaptive dipole array. *IEEE Transactions on Antennas and Propagation* 52: 1262–1269.

Hui, H.T. 2002. Reducing the mutual coupling effect in adaptive nulling using a re-defined mutual impedance. *IEEE Microwave and Wireless Components Letter* 12: 178–180.

Kim, K., T.K. Sarkar, and M.S. Palma. 2002. Adaptive processing using a single snapshot for a nonuniformly spaced array in the presence of mutual coupling and near-field scatterers. *IEEE Transactions on Antennas and Propagation* 50: 582–590.

Parhizgar, N., M.A.M. Shirazi, A. Alighanbari, and A. Sheikhi. 2013. Adaptive nulling of a linear dipole array in the presence of mutual coupling. *International Journal of RF and Microwave Computer Aided Engineering* 24: 30–38.

Rabinovich, V., and N. Alexandrov. 2012. *Antenna arrays and automotive applications*, 206 p. New York: Springer Science and Business Media. ISBN:9781461410744.

Singh, H., and R.M. Jha. 2013. Efficacy of modified improved LMS algorithm in active cancellation of probing in phased array. *Journal of Information Assurance and Security* 8: 10 p.

Svendsen, A.S.C., and I.J. Gupta. 2012. The effect of mutual coupling on the nulling performance of adaptive antennas. *IEEE Antennas and Propagation Magazine* 54(3): 17–38.

Tennant, A., M.M. Dawoud, and A.P. Anderson. 1994. Array pattern nulling by element position perturbations using a genetic algorithm. *Electronics Letters* 30: 174–176.

Yuan, Q., Q. Chen, and K. Sawaya. 2006. Performance of adaptive array antenna with arbitrary geometry in the presence of mutual coupling. *IEEE Transactions on Antennas and Propagation* AP-54: 1991–1996.

Zhang, Y., K. Hirasawa, and K. Fujimoto. 1987. Signal bandwidth consideration of mutual coupling effects on adaptive array performance. *IEEE Transactions on Antennas and Propagation* AP-35: 337–339.

About the Book

In this book, a modified improved LMS algorithm is employed for weight adaptation of dipole array for the generation of beam pattern in multiple signal environments. In phased arrays, the generation of adapted pattern according to the signal scenario requires an efficient adaptive algorithm. The antenna array is expected to maintain sufficient gain towards each of the desired source while at the same time suppress the probing sources. This cancels the signal transmission towards each of the hostile probing sources leading to active cancellation. In the book, the performance of dipole phased array is demonstrated in terms of fast convergence, output noise power and output signal-to-interference and noise ratio. The mutual coupling effect and role of edge elements are taken into account. It is established that dipole array along with an efficient algorithm is able to maintain multilobe beamforming with accurate and deep nulls towards each probing source. This work has application to the active radar cross section (RCS) reduction. This book consists of formulation, algorithm description and result discussion on active cancellation of hostile probing sources in phased antenna array. It includes numerous illustrations demonstrating the theme of the book for different signal environments and array configurations. The concepts in this book are discussed in an easy-to-understand manner, making it suitable even for the beginners in the field of phased arrays and adaptive array processing.

© The Author(s) 2015   53
H. Singh et al., *Active Cancellation of Probing in Linear Dipole Phased Array*,
SpringerBriefs in Computational Electromagnetics,
DOI 10.1007/978-981-287-829-8

Author Index

© The Author(s) 2015
H. Singh et al., *Active Cancellation of Probing in Linear Dipole Phased Array*,
SpringerBriefs in Computational Electromagnetics,
DOI 10.1007/978-981-287-829-8

Subject Index

A

Active cancellation, 1–3, 10, 13, 49
Adapted pattern, 2, 3, 6, 7, 10, 13, 15, 18, 23, 25, 27, 30, 32, 34, 39, 42, 44, 49
Adaptive array, 1, 2, 20
Adaptive nulling, 2
Antenna impedance, 3, 20
Aperture distribution, 3
Array processing, 20, 37
Array response, 2, 6, 37

C

Convergence rate, 13, 27

D

Degrees of freedom, 2
Design parameters, 1, 3
Desired signal, 1–3, 5, 6, 8, 10
Dipole array, 2, 3, 5, 6, 10, 20–23, 25, 27, 32, 39, 42
  linear, 3, 4–6, 10, 23, 25, 27, 30
  non-uniform, 2
  parallel-in-echelon, 20, 22, 23, 30, 34, 44
  side-by-side, 20, 21, 23, 32, 39
  uniform, 4, 23
Direct-data domain algorithm, 2
Direction-of-arrival, 2, 3

E

Edge effect, 37, 39, 44, 49

F

Frequency, 1, 3

G

Genetic algorithm, 2
Geometrical configuration, 2

I

Inter-element spacing, 3, 7, 23, 25, 39, 44

M

Mainlobe, 3, 5, 8, 10, 13, 15, 18, 23, 25, 27, 30, 32, 34, 39, 42, 47
Method of moments, 2
Modified improved LMS algorithm, 5, 6, 3, 10, 13, 30, 47
Multilobe beamforming, 3
Mutual coupling, 2, 3, 7, 20, 23, 25, 27, 30, 32, 34, 39, 42, 44, 47
Mutual impedance, 2, 3, 20–22, 30, 39
Mutual impedance matrix, 2

N

Null, 1–3, 5, 8, 13, 15, 18, 23, 25, 27, 30, 32, 34, 37, 39, 42, 47, 49

O

Output noise power, 3, 6–8, 13, 20, 23, 25, 30, 34, 39, 44

P

Power level, 1, 8, 10
Probing sources, 2, 3, 5, 6, 8, 10, 13, 15, 18, 23, 25, 27, 32, 34, 39, 42, 44, 49

Q

Quiescent pattern, 3, 13, 15, 23, 39, 44, 49

R

Radiation pattern, 2, 4, 39
Received signal vector, 3, 6

© The Author(s) 2015   57
H. Singh et al., *Active Cancellation of Probing in Linear Dipole Phased Array*,
SpringerBriefs in Computational Electromagnetics,
DOI 10.1007/978-981-287-829-8